The Curse of the Labrador Duck

My Obsessive Quest to the Edge of Extinction

GLEN CHILTON

Simon & Schuster
New York London Toronto Sydney

Simon & Schuster
1230 Avenue of the Americas
New York, NY 10020

First Simon & Schuster hardcover edition September 2009

Designed by Kyoko Watanabe

Manufactured in the United States of America

1 3 5 7 9 10 8 6 4 2

Library of Congress Cataloging-in-Publication Data

Chilton, Glen.
The curse of the Labrador Duck : my obsessive quest to the edge of extinction / Glen Chilton.
p. cm.
1. Labrador Duck. 2. Adventure travel. 3. Endangered species. 4. Chilton, Glen. I. Title.

QL696.A52C473 2009
598.4'1—dc22 2009004747

ISBN 978-1-4391-0247-3
ISBN 978-1-4391-2499-4 (ebook)

Acknowledgments

I owe a great debt to many curators who agreed to give me access to their Labrador Ducks, which count among their most valued treasures, or who dug through their records to assure me that they didn't have a Labrador Duck after all. That group includes Mark Adams, David Agro, Delise Alison, Miloš Anděra, Ernst Bauernfeind, Ingrid Birker, Joe Bopp, Katrina Cook, Steve Cross, James Dean, René Dekker, Felicity Devlin, Siegfried Eck, Scott Edwards, Clemency Fisher, Michaela Forthuber, Anita Gamauf, Michel Gosselin, David Green, Andy Grilz, Ed Hack, Shana Hawrylchak, Stéphane Herbet, Janet Hinshaw, Rüdiger Holz, Norbert Höser, Shannon Kenney, Mary LeCroy, Georges Lenglet, Vladimir Loskot, Herbert Lutz, Brigitte Massonneau, Julia Matthews, Gerald Mayr, Brad Millen, Jiří Mlíkovsky, Nigel Monaghan, Bernd Nicolai, Dominique Nitka, Patrick O'Sullivan, D. Stefan Peters, Matthieu Pinette, Alison Pirie, Robert Prys-Jones, Josef H. Reichholf, Julian Reynolds, Nate Rice, Douglas Russell, Frank Steinheimer, Paul Sweet, Ray Symonds, Claire Voisin, Damien Walshe, Michael Walters, Marie-Dominique Wandhammer, Erich Weber, and David Willard.

Every writer should be so lucky as to have an agent like Rick Broadhead. My dear friends Pat and Jackie Walsh polished the burrs of the first draft. My editors Jim Gifford, Kerri Kolen, Franscois McHardy, and the rest of the crews at HarperCollins and Simon & Schuster were endlessly patient in guiding me through this publishing adventure. Jan Dohner put me on to Charles Darwin's use of the term *Labrador Duck*. Alexandra Mazzitelli found details of Audubon's original *Pied Duck* painting. Michael Duggan set me straight on Jacques Cartier's claims about God's opinion of Labrador. Mike Sorenson was a star in the genetics laboratory. Terry McLaughlin graciously filled the void on a rough day in Elmira. Captain Randall Sherman chatted happily about Theodore. Margot Morris set aside some precious eggs after a tough winter.

I speak only English, and even that is pretty shaky. Many colleagues have helped me by translating letters and documents, including Julie Rainard, Matthias Amrein, Antoine Sassine, and Samuel Schürch. Andrew Geggie provided endless help in answering my silly questions about the historical use of Canadian place-names.

Many friendships have been forged along the path to the Labrador Duck, and I am grateful for the support and wisdom of these new friends, including Chris Cokinos, Barbara and Richard Mearns, and Errol Fuller. My travel companions on these journeys have included Kathleen Chilton, Errol Fuller, Georgina Brown-Branch, Julie Rainard, Sarah Shima, Jane Caldwell, and my dear, dear wife, Lisa.

This book is dedicated to Kathleen Chilton—a woman who loves life, has no regrets, and knows how to laugh.

Contents

The Curse of the Labrador Duck

Introduction

I was a nervous and obsessive child. They say that some children suck their thumbs while still in the womb; I spent those nine months chewing my fingernails. Ten minutes after I learned to tell time, I became a habitual clock watcher. I owned the biggest dictionary in the fifth grade. As a nervous and obsessive child, part of my job was to collect things. I collected NHL hockey cards and British postage stamps and *Batman* comic books and balsa wood gliders and buttons with funny sayings and, well, you get the idea.

At that time, Brooke Bond Foods Limited, makers of Red Rose tea and Blue Ribbon coffee, took to including small trading cards in their packages. The cards were designed for children, and the sales strategy might have been to get us hooked on caffeine early on. Each year's cards had a different theme. The 1969 theme was "The Space Age," and 1971 was "Exploring the Oceans." Neil Armstrong and Jacques Cousteau were very big at the time. Given my family's British heritage, it isn't surprising that we went through a swimming pool of tea each month, and so each year I managed to get most of the cards in each set of forty-eight. Any that were missing at year's end could be purchased for two cents apiece and pasted into a twenty-four-cent collector's album.

The 1970 theme was "North American Wildlife in Danger." This was a pretty heavy topic for young children, particularly when they were already so jittery from all that coffee and tea. The late, great

Roger Tory Peterson wrote the text for the cards and album that year, and he didn't ease up on the doom-and-gloom message: "There is a finality to extinction; it is the end of the line for creatures that have taken millions of years to evolve." Give me a break! I'm only ten years old, for crying out loud! Can I get a little more sugar in this tea?

Peterson wrote that "Cards 1 to 4 show birds now extinct; we shall never see them alive again." Cards 2 and 3 were the Great Auk and the Passenger Pigeon, which most game show contestants could pick as the correct answer to a question about extinction. Card 4 was the Carolina Parakeet. You never hear much about that species, probably because humans feel collectively guilty about exterminating something so darned cute.

Occupying the prestigious position of trading card number 1 that year was the extinct Labrador Duck, with a portrait painted by Charles L. Ripper. The picture shows a handsome black-and-white male duck with brown eyes and a yellow bill, standing on one leg on a rock covered in lichen and bird droppings. His head is mainly white, but for a black stripe down the middle of his crown. Black feathers on his body are interrupted by white patches on his chest and wings. Ripper didn't include a depiction of a female Labrador Duck on the tea card, partly because the hen's mottled gray and brown plumage makes it look like any other female duck, and partly because he had only enough room to feature one bird. Ripper rendered the drake with its head drawn back on its shoulders, completely relaxed, and completely oblivious to the fact that he and all of his friends are about to be smacked in the face by the "finality of extinction."

After seventeen years of issuing cards about trees and flowers and ships and famous people, Brooke Bond Foods finally twigged that adults didn't collect little cardboard rectangles, and that children weren't supposed to drink coffee and tea. They replaced the collectors' cards with miniature porcelain figurines of animals. But that tiny little cardboard rectangle with the picture of the doomed Labrador Duck that went extinct eighty-three years before I was born made a big impression on me.

Just as chicken follows chickadee, and swallow follows sparrow, an adult life committed to the study of birds followed my childish obsessions. At the University of Manitoba I started by studying undergraduate zoology, and finished with a master's degree for my research on the foraging behavior of seabirds. The University of Calgary granted me a Ph.D. for my work on the way female sparrows respond to the songs of males. Mixed in with my studies were thousands of hours in the laboratory and in the lecture hall, teaching keen young minds about anatomy and physiology, botany and histology, ecology and conservation. My passion for biology made me a popular lecturer; a student once described me as having "dangerous levels of enthusiasm." My fastidious attention to detail brought people to my research presentations at ornithological conferences.

And then, one day, I found myself all grown up, with credentials, a proper job, and an impressive title, but not one bit less obsessive than I had been as a child. The only difference is that adults have better outlets for their obsessions. And so, twenty-five years after collecting the tea card with a picture of the extinct Labrador Duck, I began research for a comprehensive account of that species for the Birds of North America series.

BUT THE LINK between my childhood experiences and my ornithological research decades later was not as direct as all that. Somewhere in my basement I still have the tea cards, and I did write the species account of the Labrador Duck, but the two are unrelated. I can blame Brooke Bond Foods Limited for a forty-year addiction to caffeine, but I had long since forgotten about the card collection by the time I came to write about the Labrador Duck.

The truth is, I wrote the Labrador Duck species account for reasons altogether more practical. As one of the authors on the species account of the White-crowned Sparrow, the editors offered to give me the whole series for free if I wrote a second account. I certainly couldn't afford to buy the whole series, so the word *free* sounded pretty appealing. Since the White-crowned Sparrow account was just about the longest account in the series, I felt that things would more or less balance out if I chose a species about which virtually nothing was known. What better choice than a bird that went extinct almost before anyone noticed that it was alive? How long could it take to write the shortest account in the series?

When you start off as a nervous and obsessive child, you are likely to grow up to be a nervous and obsessive adult. I might have simply scribbled down all of the obvious facts about the Labrador Duck and dashed them off to the series' editors. Instead I set myself the tasks of ferreting out every detail ever known and accounting for every stuffed specimen, every bone, and every egg of every Labrador Duck in the world. After that, I figured that I would discover a previously unknown color wedged in the spectrum somewhere between yellow and orange, and go on to reunite baseball's American League and National League. Maybe after that I could tackle an international treaty banning cheese in a tube.

As I worked on the account, the stories surrounding Labrador Ducks became more and more bizarre, and I became more and more concerned about getting it all correct. Even after the species account was published in 1995, I continued to go after the incomplete stories. My wife, Lisa, began to describe my behavior as being that of a dog with a bone; it didn't seem like a compliment. I told my friends strange tales of stuffed Labrador Ducks. "You should write a book," they said, perhaps to shut me up. But surely someone as dangerously obsessed as I am can only write a book about an impossible task. And so I embarked on an adventure to examine and measure every stuffed Labrador Duck specimen, no matter where it was, without exception. I was determined to see where the ducks nested (Labrador would be a good start) and where they wintered (the shallow waters around New York City). Not allowing myself to stop for a breath, I would examine every Labrador Duck egg in every museum, and visit every spot on the planet where the ducks were known to have been shot. This book is about Labrador Ducks, but it is also a story of wartime atrocities, smuggling, bastard children, the richest man in the British Empire, and America's richest murderer.

Just because I am nervous and obsessive doesn't mean that I am nuts. When I say that I set out to visit every place on Earth with some tie to Labrador Ducks, I was willing to make a few reasonable exceptions. For instance, in 1986, Redonda released a five-dollar postage stamp with an image of a Labrador Duck on it. I had no intention of going to Redonda, particularly since it is an island in the eastern Caribbean with a total area of just under half a square mile. Redonda is not world-renowned for its five-star hotels and locally brewed beer; it is completely uninhabited.

I was also in no hurry to travel to the community of Brooksville, Florida. I have absolutely no doubt that Mayor David Pugh, Vice-Mayor Frankie Burnett, and the other 7,262 residents of Brooksville are warm and decent human beings. Even so, I did not feel compelled to make the journey there just because a town planner named a street Labrador Duck Road. If you feel the need, take Sunshine Grove Road north, turn left on Hexam Road, and you will find Labrador Duck Road on your right, just after Jenny Wren Road. If you find yourself at Mountain Mockingbird Road, you have gone too far.

Despite my failure to travel to Redonda and Brooksville, my quest required me to travel 72,018 miles on airplanes, 5,461 miles on trains, 1,565 miles in private automobiles and a further 1,843 miles in rental cars, and 158 miles in taxis. Add to that 43 miles on ferries, and 1,169 miles on buses, and it adds up to a whacking great 82,257 miles, or 3.3 times around the planet at the equator. As far as I could tell at the start of my adventure, there were 54 Labrador Ducks for me to see. Presumably they would all fit on a midsize kitchen table. I suppose you could jump to the end of the book to see how close I got to seeing absolutely every one, but please don't. That would ruin the punch line.

Chapter One

John James Audubon in the Land That God Forgot

Here is some advice in case you ever decide to chuck your job in order to study birds. If you are going to be an ornithologist, choose a spot that is both exotic and remote. That way, even though you will be penniless, you can at least tell yourself that you are enjoying penury with a good view. As an ornithologist working in the wilder parts of the world, you will also have a reasonably good chance of dying in the jaws of a big scary animal, instead of in debtors' prison.

I forgot these rather simple words of wisdom about ten years ago when I set out to study the songs of Puget Sound White-crowned Sparrows. This particular subspecies, found only along the west coast of North America, really loves the urban environment. For a Puget Sound White-crowned Sparrow, the only thing better for nesting than the parking lot of a mall is the parking lot of a mall with a Kmart. It will be interesting to see if the current decline of the Kmart chain of stores results in the decline of these birds along the west coast.

On the positive side, my research in Washington State and southern British Columbia never took me more than a few hundred yards from a coffee shop. On the negative side, I faced a never-ending string of questions about the microphone and parabolic reflector I carried around. Apparently a lot of people have never seen one of these large plastic contraptions, and felt that it must be some sort of gun and blast-shield combo. In my travels in and around Vancouver and Seattle, I got to explain myself to a number of very concerned police officers, and in Sooke, B.C., used my library card as identification to keep me out of jail.

At about six o'clock one Sunday morning, while recording sparrows in a small community along the coast of northern Washington, a fellow came rushing out of his house, making a beeline for me. He looked rather impatient, and perhaps just a little ticked off.

"Thank God you're finally here!" he said.

How often does life give you a straight line like that one? "I'm sorry," I said. "I got here as fast as I could."

Now, the problem with a really good line like that is that you need some sort of follow-up. I didn't have one, and so had to admit that I had no idea what he was going on about, and that my microphone and I were just trying to record the songs of birds in and around his neighborhood. With a really disappointed look about him, he explained that he was a shortwave radio buff. He thought that I was with whatever branch of the American government deals with complaints by people whose shortwave radios had been messed up by newly erected cellular telephone towers. He pointed out the offending tower on a hillside across from his house. You really have to admire his faith that the government would send out a special agent to look into radio-wave complaints that early on a Sunday morning.

A good chunk of my research on songbirds considers cultural evolution—the changes that have occurred in their songs since someone else first recorded them. I generally travel to spots that my more senior colleagues visited some decades before. As a result, I always have a nagging feeling that I am somehow late but doing my best to get there as fast as I can. That feeling swept over me on the tenth of August, 2000, when across the continent from my work on the West Coast, the ferry MV *Apollo* took me from the community of St. Barbe on the Northern Peninsula of Newfoundland, 22 miles across the Strait of Belle Isle toward Blanc Sablon, on the border between Quebec and Labrador. One hundred and sixty-seven years and thirteen days late, my wife, Lisa, and I were traveling as fast as our ferry could manage, hot on the heels of the artist and nature enthusiast John James Audubon. We were looking for a particular hillside in the tiny coastal community of Blanc Sablon, the only reasonably certain breeding site of the extinct Labrador Duck.

Although the breeding grounds of the Labrador Duck remain a mystery, it was known to have migrated through the Canadian maritime provinces and New England en route to its wintering grounds along the Atlantic coast.

IN MORE THAN a few ways, Canada's newest province shows disturbing signs of indecisiveness. For instance, the geographical limits of the region have shifted repeatedly in the past few hundred years. At one time, the name *Labrador* was used to describe honking great bits of the North, including tracts of land that now belong to Greenland and the United States. After a bit of swelling and hemorrhaging, the boundaries seem to have settled down a bit. In the 1940s, as it became increasingly apparent that Newfoundland should move on from its status as a British colony, those in favor of joining Canada as a province narrowly outvoted the group claiming that American statehood was the only way to go. As further evidence of the province's split personality, when it came time to decide which time zone the province should be in, no one could decide whether they should share a clock with Greenland or the rest of Atlantic Canada. In the end, they split the difference, leaving Newfoundland thirty minutes out of sync with the rest of the world. (Others have said that the province is something like thirty years out of sync.) Until recently, the province was known simply as Newfoundland, after the more or less triangular island that is home to the better chunk of the populace. This was rather strange, as the province has a second half, also roughly triangular. The more sparsely populated "Labrador" bit is attached to the province of Quebec on the mainland. In a show of uncustomary unanimity, the region now goes by the rather unwieldy name of The Province of Newfoundland and Labrador.

When the explorer Jacques Cartier first got a good look at the coast of Labrador in 1534, he described the place in less than glowing terms. In fact, he was downright insulting. Having stopped at spots all along the Strait of Belle Isle, separating insular Newfoundland from mainland Labrador, he claimed that the area was nothing more than "stones and rocks, frightful and rough . . . for in all the coasts of the north I did not see a cartload of earth." If that wasn't mean-spirited enough, he went on to describe the area as "the land that God gave to Cain." Other authorities claim that Cartier described the region as "the land that God forgot." After being bitten by a few thousand blackflies, he may have felt the need to insult the region with both descriptions. I have it on good authority that the Christian Bible doesn't contain any reference to God's forgetting a plot of land, but Cartier clearly felt the region was lacking in certain creature comforts, and only a biblical reference would serve to describe his irritation, even if he had to make one up. Insect repellant might have gone a long way toward cheering Cartier up, because whatever else He may have forgotten, God certainly remembered to provide the area with lots of biting flies.

Fast-forward 299 years. Audubon, now forty-eight years of age, was no stranger to a little indecision and deception himself. He was not above misleading people about his country of birth, his lineage, and even his real name. I suspect that he felt the need to get away from his creditors for a while, having long since given up all hope of a real income in order to study and paint birds. Writing about Audubon's journeys of this sort, biographer Ben Forkner wrote: "He had failed in every business venture he had undertaken. He was leaving behind not only his wife and sons until he could support them, but also a twisted trail of doubts and debts." Of course, when I say that it was Audubon's goal to "study and paint" birds in Labrador, I mean that he planned to shoot a lot of birds, bring their corpses back to camp, stick wires up their bums to hold them in place, twist them into postures they never could have attained in life, and *then* paint them. You don't have to look at many Audubon paintings to get a sense of what I mean.

Among the birds that Audubon was keen to find, shoot, and paint was the Labrador Duck, known at the time as the Pied Duck because of the drake's black and white feathers. The bird would have been something of a novelty in 1833, having been formally described just forty-four years earlier by the German naturalist Johann Friedrich Gmelin. His description, in Latin, was accompanied by its first scientific name, *Anas labradoria*. This publication was followed three years later by the first-ever depiction of a Labrador Duck in a book about arctic animals by Thomas Pennant, a Welsh naturalist and antiques aficionado. Pennant briefly described the drake's appearance; a white head and neck, with black back and belly. He then described the hen as mottled brown, white, dusky, and ash-colored.

Audubon showed incredible optimism in his desire to shoot a Labrador Duck on its breeding grounds in Labrador. They must have nested somewhere, but Audubon had no better idea of exactly where this might be than anyone else. The stuffed specimens he had seen had all been shot on the duck's wintering grounds between Nova Scotia and Chesapeake Bay. As a sea duck, it probably nested coastally, but which coast? Could it have been as far south as the Gulf of St. Lawrence or as far north as Quebec's Ungava Peninsula? No one knew. The vast coastline of Labrador was largely unexplored by naturalists, but Audubon was a man who kept his eyes open.

Created in 1792, this is the first depiction of a Labrador Duck. Less than a hundred years later, the duck was extinct.

On his journey to Labrador, Audubon took along his twenty-year-old son, John Woodhouse Audubon, and four of his son's friends. They departed Eastport, Maine, on June 4, on board a 106-ton schooner, *The Ripley*, having taken on two extra sailors and a "lad." John Junior and his friends probably got the shock of their lives the first time John Senior woke them up at 4:00 to go out and shoot birds, particularly since the younger Audubon apparently had to overcome what his father felt was a nasty habit of sleeping late. John Senior remained back at camp in order to paint all day long. Not that this was his way of getting out of work. In his personal journal of his travels in Labrador, Audubon wrote: "I have been drawing so constantly, often seventeen hours a day, that the weariness of my body at night has been unprecedented." The party returned to Eastport on August 31. Audubon had given himself three months without having to open a mailbox full of nasty letters laden with words like PAST DUE and vague references to collection agencies. The trip did, however, leave him a further $1,500 in debt.

Luckily for us, Audubon's personal journal of his trip to and from Labrador was particularly detailed, as it was for most of his major excursions. Unlike his other writings, his travel journals seem to have been a more intimate record of his journeys, meant to be read by members of his family, rather than an entertaining documentary for the public. Like Newfoundland and Labrador itself, he generally seems to have been of two minds. On days when the sun shone, and the blackflies were not too insistent, then everything about Labrador was a joy. "A beautiful day for Labrador," he wrote on the second of July. "Went on shore, and was most pleased with what I saw. The country, so wild and grand, is of itself enough to interest any one." However, when the roll of the sea brought a greenish hue to the faces of everyone onboard *The Ripley*, or when the blackflies made one afraid to go ashore, then Labrador was a place too vile even for Cain. They experienced "rainy, dirty weather" just six days later and "John and party returned cold, wet, and hungry. Shot nothing, camp disagreeable."

After several more days of rough seas, at daylight on July 26, *The Ripley* sailed into Bras d'Or Harbour, and moored snugly. You won't find Bras d'Or on any recent map of the region. In its place, you will find a tiny community called Brador.*

Although now little more than a spot on the map, in Audubon's time the community of Bras d'Or was "the grand rendezvous of almost all the fishermen that resort to this coast for cod." Audubon was caught completely off guard by the feverish activity of Bras d'Or after the desolation he and his company had encountered in previous weeks. Although a storm continued to bash the coast through July 27, *The Ripley* set sail from Bras d'Or for the four-mile run up the coast to Blanc Sablon.

Audubon's image of the Labrador Duck, known at that time as the Pied Duck, is not one of his most frequently reproduced images. It shows a hen and a drake on a hillside with an ocean view. The drake

* It took Andrew Geggie, a toponymist at Natural Resources Canada, to point out the link between the two communities. Geggie was also able to correct me when I tried to use my elementary-school French lessons to translate Bras d'Or into "arm of gold." Somehow I had missed the rather obvious link between the names *Bras d'Or* and *Labrador.* Geggie explained that the name of the region was originally applied to coastal Greenland early in the sixteenth century by the Portuguese explorer João Fernandes, who was a landowner and cultivator in the Azores. "Cultivator" apparently translated into Portugese as "Llavrador." João used his title of cultivator as a surname, and felt the need to apply it to anything that he discovered, even though he didn't actually discover Greenland.

is engaged in modern interpretive dance, and is in line for a very bad review in the morning newspapers. The hen stands on a nearby rock laughing. The painting's background looks much like Cartier's description of rough and barren coastal Labrador, and if you squint and use your imagination, you can see *The Ripley* bobbing just offshore. Audubon knew that books of his paintings would sell rather better with some marginal notes about his adventures with each species. In his sixth volume of *The Birds of North America from Drawings Made in the United States and their Territories*, in the notes accompanying his painting of the Pied Duck, he wrote: "Although none of this species occurred to me when I was in Labrador, my son, John Woodhouse, and the young friends who accompanied him on the 28th of July 1833, to Blanc Sablon, found placed on the top of low tangled fir-bushes, several deserted nests, which from the report of the English clerk of the fishing village there, we learned belonged to the Pied Duck."

Not one of his best-known depictions, this painting by Audubon represents Labrador Ducks on their breeding grounds in northern Canada. The rendition is based on stuffed specimens now housed at the Smithsonian.

Audubon followed up this introduction to the area with a description of the nests, claiming that they were similar to eider nests, formed of twigs and dried grass and lined with down. Since they were already deserted at the end of July, Audubon concluded that Labrador Ducks must breed earlier in the season than other sea ducks.

And so there you have it. Written years after his adventures in Labrador, Audubon provides us with the only written description of the nests of Labrador Ducks. But you have to ask, if only because I told you to, is the account that accompanied his painting sufficient evidence that Labrador Ducks actually bred in and around Blanc Sablon? It seemed like a pretty good time to make a journey to Canada's east coast.

I WAS SCHEDULED to present a paper at an ornithological conference in St. John's, the capital of Newfoundland. I managed to convince Lisa that we really needed to take an extra week of holiday time, spend a lot of money on a rental car, drive around Newfoundland, take a ferry across the Strait of Belle Isles, and eventually wind up in Labrador. "After all, honey, when are we next going to have the opportunity to be eaten alive by blackflies in Labrador?" or words to that effect. For a couple of prairie kids, this was the perfect opportunity to see lighthouses, puffins, whales, and icebergs, and to cross paths with John James Audubon, who provided us with the only description of Labrador Duck nests.

At first glance, the region around Blanc Sablon is something of a moonscape. Cartier wasn't far off in describing it as frightful and rough. If you can't feel at home without a lush garden, you should probably rethink your upcoming move to Labrador. The hills rising out of the Gulf of St. Lawrence support a few shrubs, but there isn't a tree in sight. This tundra-like landscape is particularly odd because Blanc Sablon is almost exactly as far north of the equator as London, and is farther south than both Berlin and Warsaw. Much of the coastline in this part of Labrador has sandy beaches, but you would have to choose your day at the seaside with caution, to avoid exsanguination by blackflies. These biting demons are not only abundant but also sneaky, creeping up to your hairline to deliver a painless bite that bleeds long after the culprit has left. Once the flies have died off in the autumn, it is easy to imagine the wet winter winds blowing in from the gulf without a twig to stop them, chilling any creature with a heart. People may have originally settled the area because of cod fishing, but with that fishery nearly kaput, you might ask yourself why anyone would stay.

But then you would recognize the primitive beauty of the region. Without any mountains or trees to muck up the view, you can see everything from anywhere, and there really is a lot to see. Any point where the land meets the sea makes for a pretty vista, but this is a place where stark becomes bleak in the most beautiful way—a tangle of rock and brush and wind and ocean. Labrador provides enough room for each person to be an individual, while forcing each individual to be part of the greater community. Labrador has an area 19,000 square miles greater than Great Britain, but compared to the 60 million people living in the United Kingdom, there are fewer than 28,000 in Labrador. If you are willing to brave the bugs and wander along a beach, you will probably have it to yourself, but when you walk into the bar, you will find that everyone there is a friend. There is no need to enclose your yard with a fence; if your dog runs away, one of your neighbors will bring it back. When you are tired of your own company, you shouldn't have much trouble finding someone with an accordion to play "Lights over Labrador." In May, every house has an unobstructed view of the icebergs in the gulf, and everyone can complain with a single voice about the unreliable ferry service between Labrador and Newfoundland.

STATISTICS CANADA HAS quite a lot to say about life in Blanc Sablon. Canada's central statistical agency notes the population of the area as 1,235 persons. Of these, 865 speak English and 345 speak French, but only 35 speak both French and English. You would think that this would make ordering a beer in Blanc Sablon really confusing. Of the 35 bilingual persons in the community, 20 are men and 10 are women. The other 5 must be some form of life that StatsCan considers neither male nor female. Of the 1,235 folks living in and around Blanc Sablon, 10 are members of the aboriginal population, 10 are immigrants, and 1,235 are Canadian-born nonaboriginals. At this point, I am going to guess some sort of rounding error. In the five years between two recent censuses, the population of Labrador has fallen from 29,554 to 27,864. This is fewer people than attend the average Major League Baseball game.

My wife and I got to meet a few of the more or less 1,235 residents of Blanc Sablon. We asked almost everyone we met if they had ever heard of the Labrador Duck. Most folks seemed to think that we were pulling their leg. Some tried to pretend they had heard of it when clearly they hadn't, and one fellow thought it was the name of the local junior high school hockey team. We were absolutely delighted when a waitress at The Anchor restaurant in Port au Choix was able to tell us, "They're not around anymore. They're gone. They used to breed on the peninsula." Now, doesn't that seem to be the sort of thing that every schoolchild in Newfoundland and Labrador should be taught?

Two local gentlemen who got more than the usual grilling from us were Murray Letto and Robert Plouffe, wildlife conservation officers operating out of the offices of the Conservation de la Faune, Loisir, Chasses et Pêche Québec, Environnement et Faune, Québec. At first they were a little confused about what we actually wanted. After we explained that we were looking for information on the Labrador Duck, Plouffe explained that a local hunter had shot one the previous summer and, as evidence in possible court proceedings, they had its body in their freezer. I looked for signs that I was being teased but saw none. Allowing my heart to skip just a few beats, I settled back down and told myself that the Labrador Duck has been well and truly gone for about 125 years. Sadly, I wasn't about to become world famous for discovering a breeding population of a species thought to be long extinct. I wasn't going to get my picture on the cover of *Weekly World News* next to a photo of an insect-eating baby with a face like a bat. With the help of Godfrey's *Birds of Canada* on their shelf, we were able to figure out that the bird in the freezer was not a Labrador Duck but a Harlequin Duck, quite rare in that part of the country though not yet extinct. Certainly not fair game for hunters. If I had paid more attention in French classes in school, this sort of problem probably wouldn't pop up.

"How do you do research on an extinct duck?" they asked.

"You go to places where they were thought to breed, and ask a lot of foolish questions," I explained. Since we don't know of any other breeding sites, the local geography could tell us a lot about their breeding biology. Letto and Plouffe were able to provide us with no end of worthwhile information related to Labrador Ducks and the community of Blanc Sablon. For instance, *sablon* more or less translates from French into English as "fine sand," and so the area is named after the fine white sand along the shoreline. They also told us that the bay around Blanc Sablon is quite shallow in places, which would have suited Labrador Ducks as they dived to find small shellfish. However, the bottom of these shoals is neither sandy nor muddy, but rocky. This is strange, because reports of the Labrador Duck on its wintering grounds had it feeding over sand shoals rather than rocky bottoms, hence one of its earlier names, Sand Shoal Duck.

Now it was time to look for the actual spot where young John Woodhouse Audubon apparently made his discovery of Labrador Duck nests. Where does one start, with absolutely no specific description? It really shouldn't be that difficult. After all, I have been a field ornithologist longer than Audubon's son had been alive at that point. I just had to figure out where I would I go at four o'clock in the morning if I were trying to get the lay of the land, trying to pick out spots to blast away at birds, with a minimum of tromping over difficult terrain. Even if I picked the wrong hillside, who is going to be able to tell me I'm wrong?

The community of Blanc Sablon is long and skinny, with houses spread out along the coast so that everyone gets a great view of the Gulf of St. Lawrence. The land comes to something of a 90-degree angle about halfway through the community. A hill above this point is the highest around and affords the best view of both the gulf and the surrounding countryside. The hill is also covered by the sort of low tangled fir-bushes described by Audubon. Of course, the whole damned region is covered by low tangled fir-bushes, but Audubon didn't let facts get in the way of a good story, and neither will I.

In 1833, it probably would have been a considerable slog up this hill for John Junior and his friends, particularly through the thigh-deep bushes. Since then the good and pious citizens of Blanc Sablon have made the trip a lot easier by constructing and maintaining a staircase. At the top of the hill is a shrine to the Blessed Virgin, and

AVE, a prayer to Mary, spelled out in letters big enough to be seen from the Vatican. Toward the bottom of the hill, below the shrine, must be the world's largest and most nautical rosary, with the five groups of ten beads represented by fishing floats in different colors, connected by the type of thick nylon rope used to tie ships to a dock. If this rosary isn't in *Guinness World Records* as the world's largest, it should be.

In his narrative accompanying the painting of the Labrador (Pied) Duck, Audubon claims to have been told in 1833 by the English clerk of the settlement that the nests on the hillside in Blanc Sablon had been constructed by Pied Ducks. I had not been able to find the name of the clerk but did discover that he probably didn't hold onto his job for too much longer. Up to that time, the region was occupied only seasonally, because the cod fishery was profitable only seasonally. According to a plaque along the road between Blanc Sablon and Brador—surrounded by swarms of ravenous blackflies—in 1834, "Charles Dickers bought the rights to the Longue-Point-de-Blanc-Sablon fishing port which helped to establish the first year-round settlement." If I had just purchased a fishing port, I would probably have installed my own management team, tossing the clerk used by the previous owners. Blanc Sablon took more than a century and a half to grow from five families in 1834 to the 1,235 (or so) men, women, children and others who live there today.

So the story of John Woodhouse Audubon finding Labrador Duck nests in Blanc Sablon sounds just great, but on very close examination, bits of the story start to fall apart. The story about the Labrador Duck nests that accompanies the painting was written many years after the voyage. We can contrast that story with Audubon's 1833 personal journal of his time in Labrador, which was generally very explicit. For instance, he recorded a partial lunar eclipse at 19:30 on July 1, an observation substantiated by the good folks at the NASA/ Goddard Space Flight Center, with their massive great computers and too much spare time. If Audubon noted his observations on something as nonbirdy as a lunar eclipse in his journal, surely he would have mentioned something really important like a Labrador Duck nest.

Audubon's Labrador journal goes on to describe, in great detail, the nests and eggs of all manner of birds, including eider ducks, gulls, and terns, and yet he is strangely silent on the topic of the nests of Labrador Ducks. Indeed, on July 28, the day that his son was supposed to have found Labrador Duck nests, Audubon wrote in his journal: "The Pied Duck breeds here on top of the low bushes, but the season is so far advanced we have not found its nest."

In discussing Audubon's journals, Ben Forkner describes the author's prose as "a compressed mixture of diverse elements roughly made up of part recollection, part direct report, part hearsay, part invention, part theater, and part the imperial imagination's demand for the completed pattern." There are times when Audubon mixed parts of two lackluster stories into one really compelling one. Even though the fundamentals of Audubon's stories are probably correct, we can't take the details of his prose as reliable historical documents, particularly when two accounts of the same event contradict one another.

And so Audubon's daily journal explicitly claims that he and his party did not find Labrador Duck nests in Blanc Sablon, but his narrative that accompanies his painting claims that they did. Which are you supposed to believe? Sadly, I am forced to conclude that neither Audubon nor his son ever saw a Labrador Duck nest, and in writing the account to accompany his painting of the duck, he relied on his memory rather than his field notes. This is a lot more polite than saying he made the whole damned thing up.

LIKE ME, AUDUBON appears to have had a special affinity for White-crowned Sparrows. While in Labrador, he described them as "tolerably abundant," with a "sonorous note reaching the ear ever and anon." However, when it came to killing them, Audubon and his party were heartless recidivists. His Labrador journal tells us that they shot White-crowned Sparrows on June 24, 26, 29, July 2, 6, 15, 16, 17 . . . you get the idea. Trust me—every White-crowned Sparrow looks pretty much like every other White-crowned Sparrow. If Audubon needed to paint a representative of the species, one corpse would have done.

By the time Audubon left Labrador, his opinion of the region wasn't much better than Jacques Cartier's. He wrote: "Seldom in my life have I left a country with as little regret as I do this." I suppose you can only contribute so much blood to the local insects, and turn so much of your partially digested lunch overboard in rough seas before you come to really hate a place. However, I don't think blackflies and rolling seas were the worst things about Labrador for Audubon. His journal contains one description after another of the rape of the land by men from far and wide. He describes seabirds being killed in the hundreds and cut up for fishing bait, and 40,000 seabird eggs being collected for market by a party of just four men. He was clearly very concerned about the future of the region: "Labrador must shortly be de-peopled, not only of aboriginal man, but of all else having life, owing to man's cupidity. When no more fish, no more game, no more birds exist on her hills, along her coasts, and in her rivers, then she will be abandoned and deserted like a worn-out field."

As Lisa and I were preparing to leave the Blessed Virgin on Audubon's hillside in Blanc Sablon, we spotted a male White-crowned Sparrow. It being August, his breeding season was over, and he was foraging frantically, trying to fatten up before starting his migration south. And then my little bird did something unexpected for that time of year. He sang. He gave us just one song, and then fell silent. I felt as though he were saying, "Good-bye, and thanks for coming."

At that point, I wish that I could have been in contact with Audubon's spirit, just to give him an update on Labrador. The harvest of wildlife continues, but with a far greater sense of our responsibility. The human population of Labrador is gradually dwindling, but the remaining people show no evidence of a mass exodus. Just four decades after Audubon's visit, the Labrador Duck earned the dubious distinction of becoming the first species of bird endemic to North America to be driven to extinction. The last individual collected in Canada was shot in 1871. The very last bird that we know of was taken off Long Island in 1875. However, most of the plant and animal species seen by Audubon in Labrador are still there. I don't think that God has forgotten about Labrador. He probably just expects us to take better care of it.

Audubon and his party hadn't found Labrador Duck nests, but someone, somewhere, must have, because I was off to see some eggs.

Chapter Two

Scotland in a Day

Having had no success with Labrador Duck nests, I decided to try searching for their eggs. There certainly weren't going to be a lot of them to examine in the world's museums. Indeed, the actual number might be as not-very-high as nine, or as really-blessedly-low as none whatsoever. The trouble here is that, while you can look at a stuffed bird and say "yes it is" or "no it isn't" a Labrador Duck, one duck egg looks pretty much like any other duck egg. They are more or less round in cross-section, sort of oval in the other plane, a bit more pointy at one end, and rather more blunt at the other. When you come right down to it, eggs are egg-shaped.

The reason I say the number of remaining Labrador Duck eggs might be as few as none whatsoever is that all nine purported eggs were collected before the scientific fanaticism for scrupulous record keeping was established. An egg may even have the words *Labrador Duck* printed on the side, but that is no guarantee that the collector wasn't referring to a breed of domestic duck known by that name 150 years ago. Luckily, we live in an age full of nifty scientific techniques, including the ability to analyze very small bits of genetic material. Such scientific wizardry will usually give us an answer we can rely on. And so I set out to visit each one of the nine eggs with some claim to being produced by a Labrador Duck, and then to use genetic trickery to find out one way or the other.

Keep in mind that when I say I wanted to examine eggs, I mean that I wanted to see the shells of the eggs. An intact egg, left on a kitchen counter top, will eventually rot and explode. This leaves a gucky mess that no one really wants in an otherwise orderly and odor-free collection of natural history artifacts. And so the collector must make one or two small holes in the shell through which the yolk and the white can be blown out.

If you think back to your last efforts to make an omelet, you might remember that some thin membranes remained behind, attached to the shell, after you dropped the contents into a mixing bowl. These membranes were deposited around the yolk and white by the hen's reproductive tract before she started to produce the shell. These membranes contain cells produced by the hen, and therefore also contain genetic material. Hopefully the genetic material has remained sufficiently intact to figure out which species of bird produced it.

MICHAEL WALTERS MUST get a lot of invitations to play Father Christmas at children's parties. He has a beautiful white beard, a florid complexion, and a Santa-like physique, and when I first met him he was keeping his trousers up with bright red suspenders. The advertising executives at Coca-Cola might have had Walters in mind when they created their image of Santa Claus.

For the other eleven months of the year, Walters was curator of the largest bird egg collection in the world at the Natural History Museum in Tring, England, a position he has held since 1970. According to Walters, the museum's collection contains between 1 million and 2 million eggs, but he hasn't yet got around to counting them. This is a man who clearly loves eggs and his job caring for them.

Walters and I, some years earlier, began a correspondence concerning the eggs in his care that may, or may not, be the products of Labrador Ducks. He sent me the known history for three of his eggs. To me, the details of two eggs looked very similar, and after double-checking, Walters found that the Tring collection had only two possible Labrador Duck eggs, not three. When you are in charge of a couple of million eggs, this falls within an understandable margin of error.

The history of the museum's first Labrador Duck egg is reasonably straightforward. It was one of many specimens in a large collection purchased in 1889 by Lord Rothschild from someone named Count Roedern. Where the count got the egg, how much Lord Rothschild paid him to part with it, and whom Roedern had to bribe to get the title "Count" are details lost to history. The egg is a pale olive-buff color, and has *Canard Labrador,* inscribed in ink, and *Labrador 8 Juin,* in pencil. It has been in Tring ever since. The egg is known by the rather uninspiring number 1962.1.559. Walters measured it at 61.8 mm by 44.1 mm, and described it to me as very pale grayish-cream in color, smooth and glossy, and slightly soapy.

In the 1800s, when British financiers wanted to conduct trade in the Far East, the most appealing sea route was the Northwest Passage across the top of North America. The fact that no such passage had been found didn't dampen the enthusiasm of those who believed that such a route must exist, and there seemed to be no shortage of brave men willing to spend several years of their lives attempting to find it.

The most incredibly ill fated of these expeditions was commanded by Sir John Franklin, who tried once in 1818, and again in 1825. Too tenacious to give up, or perhaps too stupid to know when to quit, he tried again in 1845. This time he was generously equipped with two ships and a complement of 138 seamen. Of those, 33 perished during the first three horrible winters in Canada's frozen north; Franklin was one of them. For most of that period, the ships were trapped in ice. In the spring of 1848, the remaining 105 men abandoned ship and tried to walk south to safety. None of them made it.

I suppose that no sane English seaman would be willing to sail into perilous and uncharted waters unless the British Admiralty promised to send out a search party if the expedition became stranded. Hence, over a ten-year span, more than a dozen expeditions were mounted to search for survivors of the last Franklin expedition. One of these, HMS *Investigator*, under the command of Robert McClure, sailed straight into the Arctic ice packs in the fall of 1850, and soon became hopelessly trapped in the ice. After several years without any contact with the outside world, in 1855, the crew of the *Investigator* was rescued by *HMS Resolute. Investigator*, on just its second voyage, was abandoned.

Still in the 1800s, but now on the east side of the Atlantic, we find Canon H. B. Tristram amassing a huge collection of birds' eggs. Tristram acquired most of these eggs from persons of more robust constitution than his own, including his cousin Henry Piers, assistant surgeon on the soon-to-be-abandoned *Investigator*. The Natural History Museum in Tring is home to the second possible Labrador Duck egg, collected by Piers during a fruitless search for the lost Franklin expedition.

In a letter written (in the most atrocious handwriting) in 1901 (or 1907) from Torquay, Tristram related:

> *I never had much confidence in the authenticity of the Labrador Duck egg. My cousin, Piers, was an officer in Captain Collinson's (afterward Admiral Sir Richard) ship* Investigator *for the search after Sir J. Franklin. They were, as you may remember three years away. Piers brought me a number of arctic skins and eggs. Ivory Gull (but not its eggs) the Arctic Black Guillemot* [scribble, scribble, scribble], *Snow Goose, American Eider. He told me that this egg, got as they were going out their first spring, was the egg of a small black and white eider, but he got no skin of the bird. He declared when I showed him the drawing of the Labrador Duck that that was the drake of the layer of the egg . . . I never put the egg in the series when I had a collection for I mistrusted it. All I can say is that it came from the Arctic exploration, collected by an officer who was not a naturalist but a good skinner + taxidermist.*
>
> *I am here seeking a warm climate to get rid of bronchitis—I might as well have gone to Labrador.*
>
> [scribble, scribble]
> *H. B. Tristram*

So, if we are to believe Canon Tristram's memory, we can assume that the egg was collected by his cousin Henry Piers in the western Arctic in the spring of 1851 and was held safely through several arduous years before being turned over to Tristram. As measured by Walters, egg number 1901.11.15.266 is 61.5 by 43.8 mm, olive green, smooth and glossy, and slightly soapy.

Lisa and I were in England for the summer. I was studying the songs of Pied Wagtails, and she was investigating important matters of human health in the laboratories of a large pharmaceutical company. When it was time for me to examine the museum's purported Labrador Duck eggs, nearly two years after my visit to the Audubons' hillside in Labrador, Walters took me to a section of the museum with row upon row of tall steel cabinets. He opened one cabinet and peeked inside for a few moments. He then closed the cabinet and opened the one next to it. He then closed that one and opened another cabinet in a different row. This song and dance continued through a couple more cabinets, and I was beginning to wonder how long it was likely to go on, when a second curator pointed out to me that none of the cabinets had labels on them; this is apparently Walters' ingenious method of foiling any would-be thief who had designs on a particularly valuable egg. All of this makes me think that Walters' mind must be a tangled maze, something like Santa's job of remembering which little girls and boys had been good, and which were in line for lumps of coal in their stockings. Walters eventually found the correct spot, and set me up with the two putative Labrador Duck eggs in his care. I had a good look and assessed the potential for extracting material without damaging the shell.

Finding a person who could and would do the DNA analysis proved to be trickier than it should have. I had to make a return visit to the museum the following year to actually extract material from the eggs for DNA analysis. Walters had retired as curator of eggs, although he continued to work on at the museum in the capacity of research assistant. Assuming Walters' role as curator was Douglas Russell. Russell seemed disoriented when I arrived at the museum's reception desk, and I put this down to his trying to do too many things at once. Later he admitted that he had misremembered the date of my visit, and hadn't expected me for another month. To be fair, Russell really had a lot on his plate. In addition to all of the eggs at Tring, he had taken on responsibility for all of the collected birds' nests as well as the collection of pickled birds.

I started with the off-white egg from the Rothschild collection, the easier of the two. This egg has just a single hole near its equator, elliptical in shape, and large enough for me to work with. Using fine dental probes, I snagged some dried goop without upsetting the egg. The shell rattled a bit when shaken, and with great care I extracted an elliptical bit of white material that was bigger than the hole, although I had no idea what it actually was.

The green-gray egg Tristram Piers had collected in the Canadian Arctic was more of a challenge. It has two polar holes, both exceedingly small. Using magnifying lenses strapped to a headpiece, I could see that, in getting the yolk and white out, the collector had left some hairline cracks around one of the holes. Any pressure in that area would leave a great cavern, and make me very unpopular. This left me the other, very small, hole to work with. Luckily, when shaken, fluffy bits of material that looked like blown attic insulation fell out. The more I shook the egg, the more material I got. Again, without knowing exactly what the material was, I felt that it was sure to have some duck DNA in it. About ten minutes of shaking provided sufficient material, and I packed up my gear.

My little tubes of eggshell scrapings were dashed off to Michael Sorenson at Boston University. Sorenson had been working on the DNA of birds for sufficiently long that he could provide a clear-cut answer to the question of which eggs, if any, had been produced by Labrador Ducks. Just as important, if the eggs had not been produced by Labrador Ducks, Sorenson could say exactly which duck species had laid them, greatly reducing the chance that anyone would claim he had made a technical error. The study involved the use of some pretty impressive technology—all the more impressive because I didn't fully understand it.

The laboratory protocol works best when the sample is fresh. It is a lot more challenging when it has been drying inside an eggshell for more than a century, but my mother always said that if a task wasn't a challenge, it wouldn't be fun. Sorenson added enzymes to the sample to break up cell membranes, and then he removed contaminants like protein from the precious genetic material. Whether it is a study of duck eggs or a crime scene investigation, there generally isn't a lot of DNA to work with, so Sorenson used a technique called polymerase chain reaction to create millions of copies of segments of DNA. Then Sorenson determined the particular sequence of DNA subunits for each egg sample, allowing for a comparison with genetic material taken from the feather plucked from a stuffed Labrador Duck. It was the first time that anyone had used DNA extraction and amplification on old shells of birds to determine their identity.

Sorenson sent me the results, and it wasn't good news. DNA analysis of the material extracted from the egg purchased by Lord Rothschild from Count Roedern showed it to be the product of a Mallard, or one of the many ducks closely related to Mallards, including domestic ducks. Lord Rothschild didn't get his money's worth on that go-around.

Analysis of the material taken from the egg brought back from the Canadian high Arctic by intrepid assistant surgeon Henry Piers showed it to be the product of a Common Eider. Certainly a black-and-white seaduck, but not a Labrador Duck. Two eggs down, seven to go.

IN 1977, MICHAEL Walters received a letter from Mr. Sean O'Connell of Falkirk, Scotland. Writing on March 5, O'Connell explained that he had a collection of bird eggs, including one with the words "Labrador Duck," "Calton," and the initials "H.S." written on it. The egg was 2 7/16 inches by 1 3/4 inches and grayish white in color. The eggs had come to O'Connell fifteen years earlier as part of a big collection from a ninety-year-old gentleman whose own father had been a collector. Many of these eggs had been collected in the mid-nineteenth century. O'Connell asked if any conclusion about the validity of the egg could be drawn based on a comparison of the size of the eggs to the size of a stuffed Labrador Duck, or if the ink on the eggshell could be dated to the mid-nineteenth century. Could Calton be the name of a lake, river, or village in Canada? asked O'Connell.

Walters responded to O'Connell's letter on March 10, explaining that the egg could very well be the product of a Labrador Duck, but that techniques did not exist to support or deny this. Walters went on to describe an additional six Labrador Duck eggs in Dresden, speculating that they had probably been destroyed in World War II. He continued in the most gentlemanly way to suggest that "anything as rare as that ought really to be in an institution rather than in private hands," and explained that the egg would be gratefully accepted by either the Natural History Museum or the Royal Scottish Museum.

After waiting very patiently for seven years for a response, Walters wrote to O'Connell again, asking if he had made any progress in establishing the identity of the egg. He went on to suggest that the egg could be sent for examination to the Natural History Museum by post, if packed in a biscuit tin with plenty of cotton wool. Not altogether surprising after seven years of silence, this letter got no response, and Walters made no further attempt to contact O'Connell. With nothing but the three photocopied letters in hand, like a dog with a bone, I set out to find O'Connell and his egg.

My first letter, sent to Mr. O'Connell's address as of 1977, received no response. I then dashed a letter off to the Central Scotland Chamber of Commerce in Falkirk, explaining that I was searching for O'Connell and his Labrador Duck egg. They leapt straight to the task. Executive Director Ken Whamond wrote to say that their initial inquiries had failed to find O'Connell, but that they were willing to continue searching if I wished. They also sent me a copy of a letter from Bob McGowan of the National Museums of Scotland in Edinburgh and a newspaper report from the *Glasgow Herald.* McGowan's letter said: "I think that the chance that Mr. O'Connell ever had a genuine Labrador Duck egg is so remote that it is quite impossible. Please find attached a copy of the press cutting which refers to his conviction." His conviction?

The article from the *Glasgow Herald* of Tuesday, August 30, 1977, was entitled "I'm Smashing All My Eggs Says Birdman: Nest Robber Fined £150." The newspaper's staff writers had provided the gory details of the heinous crime. According to the article, O'Connell, then fifty-six years old, had been arrested and charged with having taken eggs from the nests of protected birds. Members of the Royal Society for the Protection of Birds had tipped off authorities. The sheriff had confiscated the eggs in question, and ordered O'Connell to pay the £150 fine at a rate of £5 per week. In response, O'Connell had smashed the remainder of his collection, which he had intended to pass along eventually to museums.

At this point I began screaming. The thought that O'Connell

might have destroyed a Labrador Duck egg, one of the very few in existence, perhaps the *only* one in existence, made me feel ill. If it wasn't a real Labrador Duck egg, it wasn't worth tuppence ha'penny, but if it was what O'Connell thought it was, the shell of this egg is probably worth its weight in gold. I asked Mr. Whamond to continue to search for O'Connell. A month later, he sent me the good news that O'Connell was still alive and living in Falkirk, and that he continued to have a collection of eggs, which, despite what he told the newspaper, he had not destroyed. Mr. Whamond suggested that, if I were to prepare a brief, one of their consultants, Dr. Alex Thompson, would pursue the matter further. I sent off the brief, and that was the last I heard of Mr. O'Connell and his egg for another five years.

While waiting, I started looking into the possibility that the inscription *Calton* on O'Connell's egg referred to a Canadian placename. I turned to Andrew Geggie, toponymist-without-parallel at Natural Resources Canada, for help in figuring out the options. Possibility number one: there is a small community known as Calton in southern Ontario, west of Toronto and just a few miles from Lake Ontario. According to Geggie, the community was given its name by its first postmaster, Duncan McLauchlan, an immigrant Scot. However, by the time Calton, Ontario, was established, the Labrador Duck was either already extinct or very nearly so. Possibility number two: Calton Point, sometimes labeled Catton Point, is a finger of land in the Yukon Territory, pointing out into the Beaufort Sea. If you were kayaking along the Arctic coast, you would have to pass between Calton Point on the mainland and Osborn Point on Herschel Island to get through Workboat Passage. Of course, possibility number three is that the word, Calton, has nothing to do with where the egg was collected. Maybe it was the name of the collector's nephew, and the uncle needed a reminder to buy the boy a birthday card and didn't have a notepad handy.

IN THE SUMMER of 2001, Lisa and I were really ready for a vacation. I was suffering the hangover of a very heavy teaching load, and Lisa had just finished her Ph.D. We agreed that a trip would make a great graduation present; somewhere nice but not too expensive. Somewhere with friendly people, worthwhile museums, notable architecture, and really good Scotch whisky. Somewhere like Scotland.

You have to give me credit for being reasonably clever, convincing my wife that her graduation gift should be an opportunity for me to try to find O'Connell and his egg. In the weeks leading up to our trip, we got in touch again with Ken Whamond. He gave us the telephone number of Alex Thompson, who assured us that Mr. O'Connell, although elderly, was still alive. Thompson gave us O'Connell's telephone number and offered to alert him that we would be calling when we arrived in Scotland. I asked Thompson to emphasize my goals to Mr. O'Connell. They were: 1. to find out if O'Connell still had his Labrador Duck egg; 2. to explain what I had found out about the place-name Calton in Canada; 3. to explain that science now had a technique to determine whether his egg had been produced by a Labrador Duck or by some other duck; and 4. to visit him, examine his egg, and extract material for DNA analysis. I most certainly did not want to: 1. turn O'Connell over to the Bird Police for having eggs that he shouldn't or 2. steal O'Connell's egg.

As soon as we got settled into our hotel room in Glasgow, I wanted to ring up Mr. O'Connell, but this left me with a problem. The British telephone system completely bamboozles me. For me, the only thing worse than trying to make a call from a British telephone box is trying to make a call from a British hotel room, and so Lisa and I went in search of the lesser evil. We wedged ourselves into a telephone box on the high street, swiped a telephone card, apologized to an operator for pushing all the wrong buttons, and eventually got O'Connell on the line. At that very instant, a road repair crew started up their jackhammers just outside the telephone box.

Despite O'Connell's thick accent and the noise from the destruction of an innocent roadway, I was able to form the impression that he had been waiting for me to call, so that he could politely tell me to bugger off. He tried to ring off twice, but I used all of the best manners my mother had taught me in order to keep him on the line. I explained how excited I was to be speaking to him and how eager I was to help prove the identity of his egg. I offered to send him a copy of the Labrador Duck species account and to tell him what I had found out about Calton as a Canadian place-name. I didn't mention the newspaper article about his time in court. He told me that he would think about my suggestion to visit him the following year to examine his egg, and would get back to me. I hung up and made a rude gesture at the road repair crew.

Scots have described Falkirk as a rubbish tip; some of these were longtime residents of Falkirk. As Lisa and I walked big lazy circles through Falkirk the next day, we found it anything but rubbishy. The man in the train station's ticket booth was genuinely apologetic when he explained that there were no luggage lockers for our backpack. When we dropped in at the public library to see if the local paper had given O'Connell's court appearance any more coverage, the lady in the reference section treated us as though her whole life had been leading up to our arrival. I made myself even more popular by fixing her photocopier. Falkirk isn't on the first page of any traveler's guide to Scotland, but is a thoroughly likable place all the same.

DEAR OLD Mr. O'Connell was good to his word. He thought about my proposal to visit him to extract membranes from the duck egg in his collection. The analysis of DNA in these membranes would well and truly sort out whether or not this egg was produced by a Labrador Duck. Shortly after we arrived back in Canada, he wrote to say that we could visit him at his home the following year. That year stretched into two, but our day finally arrived.

We were stationed in Oxford in the summer of 2003 when O'Connell, now eighty-two, telephoned to invite Lisa and me to visit him at his home in Falkirk. We set a time and date the following week. We offered to take him out for lunch afterward, but he passed on the offer, explaining that he was housebound with a bad hip that was awaiting surgery.

Life is expensive. Perhaps this is a universal truth. I haven't traveled to all parts of the universe yet, but the statement certainly applies to the United Kingdom. My wage as a university professor in Canada wouldn't pay the rent on a potting shed in Britain. One of the best ways to make life even more expensive is to travel. We decided to make the trip to Falkirk as cheaply as possible, and felt that our best option was to try to do the whole thing as a day trip. And so it became a matter of pride to try to make the trip from Oxford in England to Falkirk in Scotland and back again in less than twenty-four hours.

Our twenty-four hour goal would have been a breeze in a car, but our vehicle was separated from us by the Atlantic Ocean and most of the North American continent. It is possible to travel between Oxford and Falkirk by bus at a very competitive rate, but that would require more than twelve hours in transit each way. A train can do the journey in comfort and reasonable time, but to get a rate we could afford, we would have had to book well in advance—preferably before the end of the Crimean War.

That left carjacking and air travel. Although I have a vaguely threatening look about me, Lisa looks altogether too disarming, which reduced our options to one. Oxford doesn't have an airport, and neither does Falkirk. Even so, a bargain-basement airline offered us a very good fare between London, Gatwick, and Edinburgh if we could fly at unfashionable times. It would be a little complicated, but if we squeezed hard enough, we could fit the whole journey into the allotted twenty-four hours. And so, the journey began.

We were up at midnight, just two hours after going to bed. A cab took us to the Oxford bus station, and a bus delivered us to Gatwick by 4 a.m. After a prolonged and unexplained delay on the tarmac, we were off to Edinburgh, arriving at 8:10. Another bus took us to the train station, and by 11:00 we were on our way to Falkirk. We walked into town, jammed down some lunch, and watched the noon-hour crowd in anticipation of our 2 p.m. meeting.

O'Connell welcomed us to his home, high in an apartment block with an impressive view of the surrounding system of lush parks. Awaiting hip surgery that would hopefully set him right, he slowly came forward with the aid of a walker to greet us. His hip clicked frighteningly whenever he moved, and the sudden termination of a sentence with a sharp inhalation of breath told us that we were dealing with a gentleman in considerable pain. Despite all of this, O'Connell seemed genuinely pleased to see us. He was ready for us with a kettle just off the boil, a jar of instant coffee, a can of evaporated milk, and a plate of biscuits. Lisa tucked in, but I was a little too excited to think of anything other than eggs.

O'Connell led us into the second bedroom, which housed his collection. Cabinets lined the walls, and the little boy inside me began to imagine the treasures that were housed within. He pulled out keys for two cabinets, and we spent a half hour examining tray after tray of the shells of birds of all description. Terns and warblers and bustards and owls and eagles; we saw about two thousand eggs, all set neatly in cardboard boxes with identifying tags. Each box contained all of the eggs taken from a single nest. There was one exception—the egg marked *Labrador Duck* resided in a box beside a single flamingo egg. I said a little prayer of thanks that O'Connell had fibbed when he told the newspaper that he had smashed his collection, as I judged this assemblage to be of considerable scientific merit.

While Lisa and O'Connell chatted, I sat down at a card table and assessed the potential for getting material from the blown shell of the duck egg. It had two small blowholes, one slightly larger than the other. I considered the possibility that the collector, whoever he or she was, might have been so scrupulous in removing the contents of the shell that there would be nothing left for me to extract. As I peered through my goofy-looking magnifying headpiece, armed with an assortment of lenses, I was pleased to see that there was some material adhering to the interior of the shell. Using my series of small dental probes, I worked the dried goop out. Whatever the goop actually was, it was bound to have DNA.

As careful as I was, I was hindered by lack of sleep. With just a slight misjudgment of distance, I chipped a small piece of the shell from the smaller of the two blowholes. It turned a neat hole into a more jagged figure eight. My heart fell, because even though it was only a teeny fragment, just like a chip in a precious vase, it was bloody obvious. I confessed my act of vandalism, and Mr. O'Connell was very gracious, saying, "Never mind, never mind, never mind," and quipped about cracking eggs and making omelets.

O'Connell had been kind enough to invite us into his home, and I was too polite to ask him a hundred rude questions, even though my scientist's curiosity was burning me up: "So, who is going to get your collection after you die? How much money would it take to get your Labrador Duck egg away from you? How much for the rest of your collection? Is that your whole collection, or do you have more in all of these other cabinets?" Lisa astutely noted that all of the eggs O'Connell had shown us were from unusual and exotic birds, and that he hadn't shown us a single egg from a run-of-the-mill bird. Perhaps he collected only eggs of special birds, or perhaps he wanted to show us only the really good stuff.

One question that O'Connell answered without being asked was how he felt, in an era of great environmental concerns, about having collected all of those eggs. He explained that his collection had been amassed in a different era with different priorities. Collecting natural history artifacts had been a gentleman's pursuit. Even so, O'Connell felt a degree of remorse over his collection. Lisa and I had made a new friend. We were invited to stay longer, but we didn't want to tax the hospitality of an elderly gentleman in discomfort. We were off, still hoping to be in bed within our twenty-four-hour limit.

It was a long tramp back up to the Falkirk train station, but by 4:30 p.m. we were roaming the streets of Edinburgh, looking for a quick meal. The early evening found us on the bus to the airport. Our 8:50 p.m. departure time came and went with a big blank spot on the tarmac where our airplane should have been; the airline seemed insensitive to our self-imposed twenty-four-hour time limit. We arrived at Gatwick fifteen minutes late for the 11 p.m. coach. Finally arriving at Oxford at 3 a.m., we found plenty of kebab wagons doing a brisk business, but no taxis, and so had to hoof it. We arrived home at 4 a.m., twenty-eight hours after leaving.

And so, despite having missed our target by four hours, I had all of the material needed to determine whether or not the carefully guarded egg of O'Connell was something rare and valuable or something rather more ordinary. Was I in for a long and expensive trip to see the desolate spot in the Canadian Arctic where the egg had been laid? Michael Sorenson sent me a message with the results of the genetic analysis of O'Connell's egg. Like one of the eggs in Tring, rather than being worth its weight in precious metals, the egg turned out to be the product of a Mallard. It would be a while before I could sample the last six egg possibilities.

To date, my search for the remains of Labrador Ducks had been something of a bust. I had visited the spot in Labrador where Audubon's son almost certainly didn't find Labrador Duck nests. I had scampered over Great Britain in search of eggs that hadn't been produced by Labrador Ducks. It was time to change my luck, and so I set off in search of stuffed birds. No one could say quite how many stuffed Labrador Ducks were waiting for me. Published records told me that I could expect something like fifty specimens scattered among the world's greatest natural history museums, but some of these seemed more legend than substance. One specimen was to be found here, and another one way over there. Twelve countries . . . thirteen . . . perhaps more. Given that I had a life to live and a living to earn, this was going to take some time.

Chapter Three

Old Friends and Older Ducks

To this point I had been something of an imposter, claiming to be the world's most noted authority on something that I had never actually seen. Even though I had been reading and writing about Labrador Ducks for seven years, and despite having traveled to Labrador and Europe chasing down eggs and other leads, I was part of that enormously vast majority of humans who had never actually seen any of the surviving specimens. It was time to join the small but elite band of explorers who had lost their Labrador Duck virginity.

My journey to examine every stuffed specimen in the world could have started anywhere. I could have dashed off to somewhere exotic and romantic like Paris or New York City, and yet, since the species probably bred in Canada, that seemed as good a place as any to start. More importantly, having just finished a month of fieldwork on songbirds in California and Oregon, I was trying to get over a near-fatal encounter with poison oak. Even though the vesicles on my legs had stopped weeping, somehow I just couldn't get myself in the mood to struggle with the intricacies of the public transit system in Prague or explain my quest for dead ducks to Russian customs and immigration officials—yet.

Having tricked my wife into two Labrador Duck–related trips,

I felt I might just be able to pull a similar stunt on an old university friend. I was able to hunt down Gina Brown-Branch and her husband, Steve, in Ontario. We agreed that sixteen years, fully one-fifth of a lifetime, was too long without a visit; we should try to get together for a reunion.

"Well, you know, I do have research to do in Ontario and Quebec. What do you think about a road trip to see some dead ducks?"

Gina and Steve own a magnificent old farmhouse in Cambridge, Ontario, about an hour southwest of Toronto, dating back to the first European settlement of the area. The house was added to and subtracted from over the years, and passageways were inserted and then plastered over, but it still maintains a lovely cool atmosphere on a hot summer afternoon, as though too proud an old lady to allow herself to become overheated. The yard has many impressive trees, and one grand oak in the backyard was probably well along when vast flocks of Passenger Pigeons passed through that part of the province, roosting in trees of exactly that sort.

Steve and Gina breed autos in their yard. Gina drives a tank-size 1984 Mercury, while Steve drives a living room on wheels, more commonly known as a 1988 Lincoln Town Car. These vehicles are only stand-ins while Steve works on a 1960 Gentleman's Hot Rod T-Bird for himself and a 1954 Series 62 Fleetwood Cadillac for Gina. They are also restoring a giant silver bus big enough to accommodate a high school marching band on a three-week tour of the American Midwest. An assortment of other immobile cars that don't seem to fit a theme may have just pulled into the yard for a rest, liked the view, and decided to retire there. Luckily, the yard is big enough that the neighbors haven't yet pilloried them. They have hidden most of the good stuff in a newly constructed garage just a little smaller than Boeing's aircraft assembly building.

The home's interior decor has been changed a bit since it was built. Most of its nooks and crannies are occupied by an amazing collection of automobile memorabilia, and the floor has been reinforced in places to support stacks of car magazines. Even so, visitors to the house are likely to be on their third of fourth visit before they notice the car theme. This is because most walls of the house are occupied by gigantic paintings of nude women. If I wanted to display one of these paintings in my house, I would have to raise the ceiling a yard, and hire a good divorce lawyer. The women in these paintings are altogether, completely, emphatically, undeniably undressed. They are in the downstairs living room, the upstairs lounge, the dining room, and (God help me) the guest bedroom. Most were painted by a single artist, who really, really loves to paint breasts. I found it more than a little disconcerting that some of the paintings were of Gina. Her university degree is in zoology, but she now makes her living as a nude model for artists and art students. The three of us spent a day cruising the back roads of southern Ontario, getting caught up on our lost years, before Gina and I left Steve behind and set out for Toronto in Gina's Mercury.

THE AUTOMOBILE ASSOCIATION'S guide to Ontario explains that "most visitors are likely to be pleasantly surprised by Toronto's weather." Reading between the lines, this means that foreign visitors, completely ignorant about all things Canadian, and expecting to find the city dotted by igloos on an otherwise barren and frozen landscape, will be pleasantly surprised to find that the dogsleds operated by the Toronto Transit Commission generally give igloos a wide berth. Gina and I were heading for the Royal Ontario Museum, Canada's largest museum of human culture and natural history, with more than 6 million artifacts, and my first stuffed Labrador Duck. A pretty good chunk of the story concerning this particular specimen revolves around the late Paul Hahn.

Hahn might be just about the perfect example of a biologist-wannabe. He didn't have a degree in the field, but, like so many other natural history enthusiasts, he probably dreamed of spending his life chasing sparrows through alpine meadows, pretending to be a critical analytical thinker hot on the trail of some pressing and important question that only other ornithologists would care about. Hahn made his living by selling and restoring pianos. He satisfied his biology cravings by becoming involved in the activities of the Royal Ontario Museum, and he contributed as a patron of the work done by the institution.

In the late 1950s, Hahn took on the interesting but almost certainly futile task of documenting the location and history of every stuffed specimen of the extinct Passenger Pigeon in the world. This task was futile because, unlike the handful of Labrador Duck corpses still around, thousands of stuffed Passenger Pigeons remain, and many of these are found in the most unlikely places, including the trophy cases of small-town high schools.

Hahn sent questionnaires to every museum in the world that might have stuffed Passenger Pigeons in its collection. While he was at it, he asked them about their stuffed specimens of other extinct North American species, including Great Auks, Ivory-billed Woodpeckers, Carolina Parakeets, and Eskimo Curlews. Almost as an afterthought, Hahn also asked museum curators about stuffed Labrador Ducks. In the late 1950s and early 1960s, he exchanged thousands of pieces of correspondence with museum curators, and in 1963, the results of his noble quest were published as a book entitled *Where Is That Vanished Bird? An Index to the Known Specimens of Extinct and Near Extinct North American Species*. The book is, in essence, hundreds of pages of annotated lists. No one would describe it as a riveting page-turner, but more than forty years later it is still the most complete list of its kind. Although Hahn lived to the ripe old age of eighty-seven, he didn't last quite long enough to see the book published. Instead it serves as a tribute to his energy and dedication to the task. It provided me with a list of fifty-four Labrador Ducks.

Hahn, like me, couldn't bear to throw anything away, and the paperwork that accumulated in preparing his book now resides in the Library and Archives of the Royal Ontario Museum (the ROM). The museum's head librarian and archivist was happy for me to dig through it all. Although it makes for an interesting read, the material is not well organized. Correspondence from museums in India is included in the file pertaining to Italy, and the folder for Hawaii contains material from Peruvian museums.

Today, with 134,000 bird skins in its collection, the ROM ranks somewhere around thirteenth in the world, but it is first in the world in terms of bird skeletons, with 42,500. It must have irritated Hahn that the ROM, with its large and growing collection of natural history artifacts, had neither a Great Auk nor a Labrador Duck, because he set his talents to helping the museum to get one of each. A hundred and fifty years earlier, if a museum wanted a Great Auk, it would simply let it be known that it was willing to pay top dollar for a specimen. If the price was sufficiently outrageous, some debt-ridden collector would risk a trip to a remote island off the coast of Iceland to club one. Since Great Auks and Labrador Ducks are now completely and irrevocably extinct, the task becomes a bit trickier. Many museums have resorted to the next-best thing by having a taxidermist cobble together a fake from feathers of other species. For Hahn this wasn't an acceptable solution. Clearly, the only route was to buy or trade for specimens in the hands of another institution. Apparently he didn't think to steal one.

Hahn's correspondence includes letters to museum curators that strongly hint that the ROM would be very keen to purchase a Labrador Duck or Great Auk if the institution wished to give theirs up. For instance, letters to Hahn from Professor F. Ronald Hayes of Dalhousie University in Halifax imply that Hahn had suggested a swap of Dalhousie's Labrador Duck for a pair of the ROM's Passenger Pigeons. The ROM has about 150 Passenger Pigeons, and this would have been a ridiculously one-sided trade of diamonds for glass; the folks at Dalhousie were not sufficiently gullible to fall for this trade. Hayes explained that, according to Dalhousie folklore, they had been offered $5,000 for their Labrador Duck and had turned the offer down. A handwritten note in the margin of one letter suggests that Hahn was going to try to get the vice president of Dalhousie to reconsider their position, given that the ROM has one thousand visitors for every visitor to the Museum at Dalhousie. The appeal was not successful, but the ROM did eventually get both a Great Auk and a Labrador Duck.

Somewhere around 1840, a Labrador Duck was shot by Jacob P. Giraud Jr. and mounted by a well-known New York City taxidermist, John G. Bell. Giraud collected most of his specimens in and around Long Island, and I might surmise that this Labrador Duck was shot there. In 1867 Giraud gave his collection of 800 mounted North American birds, including his Labrador Duck and Great Auk, to Vassar College in Poughkeepsie, New York, where the duck sat until 1921, when it was remounted by taxidermist George Nelson, put into a glass and brass case, and placed in the capable hands of the American Museum of Natural History in New York. The duck arrived in the Big Apple in October 1921 and was promptly locked away. The ROM purchased it and the Great Auk from Vassar, receiving them in July 1965, thus moving the Labrador Duck from a locked safe in New York to a locked safe in Toronto.

The sale was arranged by Dr. Ralph S. Palmer, formerly a professor at Vassar, but at that time New York State Zoologist. The question of selling the duck and auk was apparently a delicate matter, for Vassar had refused several previous offers. The ROM's James Baillie arranged a fund-raising campaign, and more than 200 individuals and organizations contributed. The birds came through at quite a bargain—$3,500 for the duck and $7,000 for the auk.

Gina and I were met at the ROM by Brad Millen, the museum's database technician for the bird collection, who signed us in and had us fitted with Really-Bloody-Important-Visitor-So-Pay-Attention-Dammit badges. Pleased at the prospect of sharing some of the museum's great treasures with people who would really appreciate them, Millen showed us a number of stuffed specimens of extinct birds, along with other beautiful creatures that have avoided extinction so far. While Millen and Gina continued their tour, I got down to my examination of the ROM's duck.

Labrador Duck 1

My very first Labrador Duck is also the finest specimen in the world. In part, this is a tribute to the gifted taxidermist George Nelson. Millen showed me X-ray images of the duck, which revealed an amazing array of pins and supporting wires that resulted in a very lifelike pose. Between visits by ornithologists, the duck resides in its protective case in a fireproof safe, along with other particularly precious specimens like the Vassar Great Auk, two Heath Hens, an Eskimo Curlew, and two of the museum's many Passenger Pigeons. It took me forty-five minutes to measure the bill and the wings, to make notes on the colors of the bills, eyes, and feet, and to take some photographs.

If you can't afford to buy a Labrador Duck, here's how to make one. First, obtain a reasonably cooperative medium-sized duck; a Mallard or a scaup should do nicely. Spray-paint it black. Then tip its head and breast into a bucket of white paint. (I realize that, as birds,

Labrador Ducks don't have "breasts," but it is a technical expression. Just trust me on that one.) Then paint the wings white, except for the ends of the flight feathers. The tips of flight feathers and automobile tires are black for the same reason—pigment makes them better able to resist wear. Then paint a black ring around the base of the neck, and a thin black stripe lengthwise along the top of his head. Voila! You have an adult male Labrador Duck. Females kinda got shortchanged in the beautiful plumage department, being mottled gray and brown all over. This is great for hiding from predators but not the sort of pattern that gets your picture on a stamp in Redonda. Depending on their age, the feathers of an immature male were somewhere between the color of a female and an adult male.

You can allow your more creative side to emerge when it comes time to paint the bill, because we really don't know what color they were. Feathers generally retain their color after the death of the bird, but body parts like beaks and legs tend to fade over the years. It doesn't help that some taxidermists painted the beaks and feet; they may have given those body parts the colors they had in life, or they may just have had some paint left over from bathroom renovations. The tip of the bill was probably black, and the base might have been brownish, but there might have also been some red and blue mixed in. We also don't know the color of Labrador Duck eyes, because reports of the day were contradictory. The male in Toronto has dark brown glass eyes; perhaps John G. Bell was a sufficiently good taxidermist to peep at the eyes of the corpse before he put in glass eyes of a similar color. So, now that I had lost my Labrador Duck innocence and gone through Hahn's correspondence like a smutty scientific voyeur, it was time for Gina and me to drive on in search of duck number two.

For most specimens of the Labrador Duck, we know very little of their history. In centuries past, birds were shot and stuffed more as curios than as items of scientific value, an ornithological equivalent of stamp collecting. If note were ever made of the date and location of their demise or their collector, those records may not have survived. There are a few notable exceptions, including the specimen in Toronto, and the one in the very capable hands of the Canadian Museum of Nature in Ottawa; not only is its history well documented but the story is entirely Canadian, and at least as odd as anything else in Canadian history.

In the fall of 1803, the Reverend Thomas McCulloch and his family sailed from Scotland and into the harbor at Pictou, at the north end of Nova Scotia, on Canada's east coast. Beyond his qualifications as a man of the cloth, having taken a course in medicine at Glasgow University, he was also qualified as a physician at a time when many recent immigrants were ill. McCulloch accepted the post of Minister of the Harbour for the Prince Street Church, and remained in Pictou for the next thirty-five years. McCulloch House, built for the Reverend and his family around 1806, remains a tourist attraction.

The good Reverend McCulloch is remembered today as the founder of Pictou Academy (1816), which later evolved into Dalhousie College, and then Dalhousie University in Halifax, with McCulloch as its first president, from 1838 until his death in 1843. This is all very noble, but he offset his positive qualities by shooting more than his share of Labrador Ducks in Pictou. Upon McCulloch's death, his collection, including a male and female Labrador Duck, was presented to Dalhousie College.

The trick is that one of the Labrador Ducks, the female, is not a Labrador Duck at all, but rather a Black Scoter, with its bill painted to resemble a Labrador Duck's. Hoyes Lloyd, an early Canadian ornithologist, noted the error when the birds were still at Dalhousie College, and published a short paper to this effect in November 1920. Perhaps McCulloch made a mistake, and thought that he had shot both a male and female. Perhaps he felt that the display would be more attractive with both a male and female but, lacking a female Labrador Duck, doctored a female scoter for aesthetic purposes.

By 1968, Dalhousie must have realized that it didn't have proper facilities to house so valuable a specimen and its partner, and so "loaned" them to the Canadian Museum of Nature in Ottawa, where they reside today. I suspect the curators in Ottawa are hoping that Dalhousie University will simply forget that the Labrador Duck and its Black Scoter mate ever existed, and they will get to keep them forever.

Michel Gosselin, Collection Manager of the vertebrate section at the Canadian Museum of Nature, made the arrangements for my visit to examine the duck in their care. He explained that although the museum's postal address is in Ottawa, Ontario, the duck is housed at their facility across the Ottawa River in Aylmer, Quebec. As Gina and I followed the precise directions Gosselin had provided, we began to think that he had been having us on. The community of Aylmer thinned out until we were back out into the countryside. Horses . . . cattle . . . surely this couldn't be right. But, just a little farther down the road, we came across a small sign indicating that we had arrived. I had been expecting busloads of crazed schoolchildren, a gift shop, and entertaining interactive displays, not realizing that this was the national museum's facility for storing its tremendous collection of natural history artifacts, and not a museum open to the public.

Gosselin welcomed us. Like so many other curators, this is a man who clearly loves his job. He is proud of the newly constructed facility and the important work it does, and loves to have visitors who also appreciate the value of the work done there. Before bringing out the specimens, Gosselin showed us around parts of the facility that houses birds and mammals. He showed us how a handle could be turned to move the collection cabinets around, and so conserve space by not wasting any on aisles. He showed us the extraordinary measures in place to prevent pest infestation and dust, two deadly archenemies of museum collections. Security was also a primary concern, with swipe cards necessary to get from anywhere to anywhere else. Frankly, all of this security might be a bit over the top. The place is so big, and all the doors and corridors look exactly the same as all the other doors and corridors; if someone broke in one night, they would eventually grow weary of trying to find their way out and just sit down and wait for the police.

The facility reminded me of a very exclusive casino, where people are invited to lose huge sums of cash as quickly as possible. Not just anyone is allowed into a ritzy casino—you have to have lots and lots of money to lose. It wasn't that the museum was trying to keep people away from the artifacts—you just had to have lots and lots of the right sort of credentials. If Cameron Diaz, to choose a person at random, were to show up and ask to see a stuffed Whooping Crane, she would be thanked for her contributions to American cinema, and then politely but firmly turned away, with the explanation that it just wasn't that sort of facility. Because I had an appointment and a good reason to be there, and because I have "Dr." in front of my name and Cameron Diaz doesn't, Gina and I were not only admitted, but treated as honored guests.

Labrador Duck 2

The male Labrador Duck and the female scoter are in pretty good shape, given how long they have been dead. Their bills and feet are a little beaten up, and someone had painted the bills black, mustard yellow, and baby blue. The drake has a small orange-yellow patch on either wing, which, according to X-ray images, is the result of rusting wires that keep the duck upright and in more or less the right shape. They are mounted on a base covered with pebbles and dried algae, perhaps to represent the seashore at Pictou. The base is cracked and perforated in spots with exposed nails, but then the Dead Sea scrolls aren't in perfect shape either.

In fact, there are only two really peculiar things about this Labrador Duck and his unlikely bride. First, the bottom of the wooden base on which they are mounted has a crown symbol and ROYAL YEAST CAKES in large black letters, showing that it was made out of an old packing crate. Second, the drake's left glass eye is dark brown, but his right eye is lime-yellow-green. Was the taxidermist a little drunk the day he stuffed it, or had he just run out of brown eyes halfway through? When they aren't being gawked at by an ornithologist, these ducks live in a large locked gray cabinet with a Heath Hen and a small flock of Passenger Pigeons.

At one point I needed to take a toilet break, and Gosselin had to escort me out of the collections, all the way back to the reception area. The collections have no bathrooms because bathrooms require water, and water pipes come with the risk that one might burst, and water is another danger of incredible destructive potential to a natural history collection. With duck number two behind us, it was time to be on our way to Montreal to see duck number three.

• • •

When people first arrived in Canada, about 14,000 years ago, they apparently managed the crossing from Asia at a narrow land bridge across what is now the Bering Sea. As a paleontologist explained it to me, an ice age tied up so much water in the formation of ice sheets that it lowered the level of the ocean to the point that a land bridge was exposed. It sounds a little far-fetched to me, but I wasn't there at the time, so I'll have to take his word for it. These explorers went on to explore and colonize the continent, rather quickly, from west to east.

In contrast, when Europeans arrived in Canada, some 13,500 years later, they also ran roughshod over the continent, but this time from east to west. This means that historians concerning themselves with European settlement and its fallout have a lot more to say about eastern Canada than they do about the west. Being east of center, the city of Montreal has more recorded history than most communities in Canada. For instance, the Redpath Museum of McGill University, erected in 1882, was Canada's first building designed to be a museum. It must be one of the few remaining museums in the country that do not charge for admission. The Canadian Automobile Association charitably describes the museum by saying "collections are displayed much as they were then." This appears to be a polite way of saying that the displays are dark, dusty, and boring. However, if you are strolling through that part of Montreal in a heavy rainstorm, and you can't afford a cup of coffee, then I can highly recommend it.

The Redpath Museum is very fortunate to have a Labrador Duck in its collection. In 1893, Ernest D. Wintle of Montreal was strolling through the now defunct museum of the city's Natural History Society, spotted a stuffed bird, and correctly concluded that it was an immature Labrador Duck. There is no record of who shot it, where or why, or how it got into the museum. Being a young bird, not in the bold black and white plumage of an adult drake, it had managed to evade proper identification until Wintle's sharp eye fell on it. In 1926, this stuffed duck made a very short one-way migration from the headquarters of the Natural History Society at 710 Sherbrooke Street across the street to the Redpath Museum at 859 Sherbrooke Street, where it has resided ever since.

Gina and I arrived at the Redpath Museum during a downpour. I suppose we were lucky to get to the museum at all, given that the streets of Montreal seem to have been laid out by a city planner who accidentally dropped two decimal places in calculating how many vehicles the roads would have to handle. We filed in behind a large flock of elementary schoolchildren, and they gave the place the delightful buzz that can only be produced on a day away from school. It took a few minutes for the teachers and tour guides to organize the students into manageable groups, but when they exited the foyer, we were left completely alone.

To prepare for my visit, I had traded email messages with David Green, the curator of vertebrate animals. Green is also past chair of the Committee on the Status of Endangered Wildlife in Canada, and so is no slouch when it comes to extinction. Green explained that he would be out of town on the day of my visit, but he arranged for someone to meet us and to show me their Labrador Duck.

And so, when Gina and I arrived at the Redpath, we were surprised to find that no one had ever heard of us. Gina found a reception office, and inside was an unpleasant little troll who was convinced that she owned the museum and everything in it, and felt that it was her duty to keep us away from anything valuable or interesting. Explaining that I was there to see their Labrador Duck, she responded, "Oh, I don't think you are going to be allowed to see that." I described the arrangements that I had made with David Green, and offered her my business card. She wouldn't have any of it and left to find someone else to assist her in being really unpleasant to me. Was I going to be stumped on just my third duck?

The museum's horrid little receptionist was unable to find someone else equally horrid, and instead sent us up to the office of a very pleasant young lady, Ingrid Birker. She listened to my story, and said, "Well then, I had better go and get the duck for you." She set me up in an unoccupied office with a broken photocopier, a microwave oven, and a desk.

Labrador Duck 3

The Labrador Duck at the Redpath Museum, whatever its origin, was never given the opportunity to get into much trouble. Its feathers are mainly brown and gray, much like the plumage of a female, but when it is tilted just right, a slightly darker brown ring of feathers around its neck and a slightly darker stripe along the top of its head show that it once had high hopes of growing into an adult male. When it was first mounted, it probably wasn't in the best shape, being full of bullet holes and indignation. One hundred and ten or so years in Montreal museums hasn't helped its condition. The wire through the left leg that holds the specimen erect has broken through the skin in back. The webs between the toes are perforated. The feathers are generally messed up and in need of a thorough cleaning. There is an odd dark patch on the feathers of its belly, probably a grease stain from the time when the guts were pulled out. Its feet are nailed to a block of wood, 15 by 15 by 2 cm, with a sloppy gray paint job and an undercoat of blue showing through chips in the gray.

The taxidermist had given it a yellow glass eye on the left side, but didn't bother to give it a right eye, so that stuffing pokes out through

The immature drake at the Redpath Museum in Montreal stared at me with a look of reproach.

the orbit, rather like a teddy bear that has been loved too much and repaired too little. The skin around the left side of the head wasn't set in place properly, leaving its jaundiced eye to protrude unnaturally from its head, staring backward. This gave the duck a most uncanny expression. While measuring its left wing, I glanced at its head, and could swear that it was giving me a reproachful look. I wanted to explain that he had been dead for at least a hundred years before I was even born, and that I am a vegetarian, and don't shoot ducks or any other animals, and that I feed bread crumbs to ducks every chance I get, and . . . but I didn't feel that he would have been satisfied with any answer I gave. I finished my measurements, took some photographs, and let Ingrid know that I was done.

Museums face a dilemma. Usually, the most treasured artifacts are not on public display but protected behind locked doors. But a specimen hidden away isn't much good for public education on the finality of extinction, hence the dilemma. Two years earlier, after considerable discussion, the curatorial staff of the Redpath Museum had decided to put their Labrador Duck on display. Under perfect storage conditions, a stuffed bird specimen should last about five hundred years before falling apart. Nothing, except extinction and bad credit, is forever. A specimen on display, exposed to light and dust and high humidity, will disintegrate much more quickly. The Redpath Museum's Labrador Duck resides in a glass cabinet in the stairwell between the second and third floors, along with a pair of Passenger Pigeons, an "Arctic" Curlew (I think they mean Eskimo Curlew), and an extinct snail. Also on display in the stairwell are an irritated gorilla and a surprised lion. (Irritated and surprised to be dead and stuffed, I suppose.)

While at the ROM, Millen had asked me how many Labrador Ducks I had seen so far. Although it felt awkward, I answered honestly that his was my first. He knew that my goal was to examine every stuffed specimen in the world, and I had to wonder if he would have given odds against me completing the task. I now had three ducks to my credit. True, I had snapped up some easy ones, but the adventure was well and truly under way. As my plane left the tarmac, I probably would have felt a little more comfortable with the quest if I knew exactly how many specimens lay ahead of me. It was time for a little duck hunting in England.

Chapter Four

Walter Gets Blackmailed

Let's be honest with ourselves—Great Britain is, for all intents and purposes, one and one-third small chunks of rock jutting out of the Atlantic Ocean, just off the coast of mainland Europe. Even so, from this entirely unlikely setting the world has taken the parliamentary system of government; a pretty good judicial system; the language of industry, commerce, and science; and the line that defines zero degrees longitude. Having provided the world with some of its best literature, most of its best rock music, and the best beer anywhere that I have been to date, Brits have also given us soccer, the world's most popular spectator sport, even though they seem to think it should be called football. On the downside, Britain has an obsession with lawn bowling, cricket, and mushy peas, an unhealthy fascination with Pete Doherty, and an almost pathological aversion to the Argentine national soccer team. Great Britain is also blessed with more than its share of stuffed Labrador Ducks.

The *Blue Guide* to travel in England, published in 1930, describes the city of Tring in fewer words than it devotes to the Roman ruins at Richborough. It gives the population of the community as 4,352 inhabitants. It also provides the names of two hotels and explains what you can expect to pay for a night's stay at each, although the

Rose and Crown may have increased its rate from three shillings and sixpence, so you might want to call ahead. The guide describes Tring as an ancient town situated at the foot of the Chiltern Hills, tells the traveler to expect a 1¾ mile tramp from the train station to the town, and tells of a baptismal register that refers to the ancestors of the American president George Washington. Before going on to attractions beyond Tring, the guide summarizes the reason for my journey. It says: "Adjoining the town is *Tring Park*, the seat of Lord Rothschild, who has stocked the deer-park with emus and rheas (visible from the footpath through the park) and has also built an admirable *Zoological Museum* (adm. free)."

The area around Tring has been occupied for at least four thousand years, but much of its more recent history has been dominated by the Rothschild family. Nathaniel Rothschild came to Tring in 1874 and was elevated to the rank of Lord in 1885. When he passed away, thirty years later, he was thought to be the richest man in the British Empire, which seems to me to be exactly the wrong time in your life to be fabulously wealthy. While still alive, he contributed to the community in many positive ways. He had slums in Tring cleared away and replaced with modern cottages. These he promptly turned over to the town council on the condition that tenants should be charged only nominal rent during their first year. He built a hospital, supported local industry, and was generally the sort of fellow whom you want living in your community. To this day, however, the locals blame Nathaniel for ensuring that the railway station was built so far out of town. But, to be fair, the Rothschilds didn't move to Tring until forty years after the station was built.

One of the more peculiar contributions of Nathaniel and Lady Rothschild was their son, Walter Rothschild, born in 1868. From an early age, Rothschild was fascinated by all things zoological, and set about collecting natural history artifacts. As a birthday gift, Rothschild was given a museum to house his collection. The museum cost £3,300 to construct, and the builders threw in a cottage for Rothschild at no additional charge. Lord and Lady Rothschild probably imagined that their son's fascination for all creatures great and small was the harmless passing fancy of a young lad. Little did they know that Walter was to become world famous for his contributions to the field of zoology, writing many hundreds of articles and books in the field, and describing five thousand new species of animals.

The museum, too, was a cracking success. When it opened in 1892, it attracted 30,000 visitors a year, which is all the more incredible when you think that most of those visitors would have arrived on foot. Admission to the Walter Rothschild Zoological Museum was free seventy-five years ago, and it is free today.

Rothschild accumulated the largest private collection of birds ever, with about 300,000 stuffed specimens, including two Labrador Ducks, and 200,000 eggs. If you are impressed by really big numbers, you will be pleased to hear that he acquired two and a quarter million butterflies and moths by hiring more than 400 professional collectors. He also purchased huge collections from other bird maniacs. And what he couldn't collect he had created. The museum contains a model of a giant moa, a flightless bird formerly found in New Zealand but driven to extinction 500 years ago. With no moa feathers to work with, the model is covered with emu feathers. With a flair for the dramatic, Rothschild brought the model to London for a meeting of the British Ornithologists' Club, its great bulk sticking out through the top of a taxi. Before he died of cancer in 1937, Rothschild arranged that his museum and its collections be given over to the British Museum.

Oddly, most of the stuffed bird specimens weren't turned over to the British Museum, having been sold to the American Museum of Natural History in the 1930s. The trick was that, although Rothschild's income was enormous, it wasn't infinite. Sooner or later something had to give. Rothschild, who never married, found himself in a spot of trouble after a series of affairs. It was bad enough that Walter had concurrent affairs with actresses, at least one of which left him with an illegitimate daughter, but he made the mistake of an indiscretion with a ruthless peeress, which left him subject to crippling blackmail demands for most of the rest of his life. In order to raise the funds to pay off the peeress without alerting his domineering mother, Rothschild had to sell the stuffed birds in his collection. Like most of the rest of us, Rothschild spent a lot of time inventing interesting ways of getting himself into trouble.

Rothschild made it a condition of his gift to the British Museum that his museum in Tring remain a center for research into all matters zoological, as it had been for many decades. When the Natural History Museum in London found itself running short of space in the late 1960s, the museum decided to move its Bird Group to Tring, freeing up space for other collections. The Natural History Museum's collection of bird skins and eggs is, without a doubt, the finest collection in the world. Moving the specimens away from the polluted air of London to the green and pleasant lands of Hertfordshire probably extended their life span immeasurably. The curatorial staff will probably last longer too.

If you wish to visit the Walter Rothschild Zoological Museum in Tring, you simply show up, leave your vehicle in the free car park, say a cheery "hello" to a lady in a glass booth, and stroll in. If, however, you wish to visit the collection of eggs and birds skins of the British Museum, housed next door, you had better have a really good reason and an appointment. I had both. Wanting to examine their two Labrador Duck specimens, I had made an appointment two months earlier and confirmed it the week before. At 9:00, the security guard told me that each of the three men I had arranged to see, Robert Prys-Jones, Michael Walters, and Mark Adams, was keeping later hours than I, but curator Frank Steinheimer would take care of me. After issuing me a visitor's pass and relieving me of my traveling case and toiletry bag for security reasons, the guard sent me off through a maze of dimly lit hallways, in search of Steinheimer.

Steinheimer set me up with a pile of paperwork. First I had to fill out a form with my name, address, and a description of the reason for my visit. Then I had to sign a document stating that I had read the safety instructions for the museum. I was not to lick any of the bird specimens, for instance, as some were preserved with cyanide.

Labrador Ducks 4 and 5

I pulled out my notebook, calipers, rulers, and magnifying lenses, and began the work of examining the beautiful drake and hen. The male has a couple of small holes in his upper bill, and his feathers are a bit dirty, but I think that a bit of grime can be expected after being denied a bath for well over 150 years. The female is missing the small hind toe on her right foot, and the outermost toe on her left foot is broken, but otherwise she is in pretty good shape for an old gal. Her head is drawn slightly back over her shoulders, giving her a demure, submissive look. The drake's bill is slightly agape, as though he is about to say something important like, "Hey, is that guy holding a gun?" Her glass eyes are green-gray-brown and his are lemon yellow.

Birds' bodies can be prepared in one of two main ways. If a specimen has been prepared as a taxidermic mount, it is meant to depict the individual as it might have been in life, much like a trophy fish or the head of a deer with a particularly impressive set of antlers. If the work has been done by a skilled taxidermist, the result can be quite stunning. All three Labrador Ducks in Canada were prepared this way.

The problem with taxidermic mounts is that they take up an awful lot of space in museum collections, require considerable time and skill to prepare, and are easily damaged. Many birds are instead prepared as torpedo-shaped study skins, and the bird is stored on its back, with its wings pressed close to the body.

A study skin is, in essence, the skin of a bird, cleaned and treated with a preservative to discourage pests; the body is stuffed to the proportions of the original specimen. A forty-year-old publication by the National Museum of Canada provides the most thorough description of how to turn a dead bird into a study skin. To save you from reading its forty-two pages of blood-soaked glory, I will summarize. Step one, don't bother; the required permits are enough to discourage anyone. Step two, it is crucial to ensure that the bird has really expired; no one wants to be embarrassed by a bird pretending to be dead. Step three, an incision is made through the skin on the belly side from the breastbone backward. Step four, working through this incision, turning the skin partially inside out, most of the body is gradually removed, taking care not to puncture anything juicy that will make a mess. The skull, minus brain and eyes, and parts of the wings and legs are left in place to provide shape. Study skins are generally not given glass eyes. Step five, do a little exploratory surgery to determine the bird's sex by identifying testes and ovaries. Step six, by this time there is bound to be guck on the feathers, and this must be cleaned off. Sawdust and cornmeal are among the best agents to help keep feathers clean. Step seven, a stick or wire is inserted the length of the body to provide support. Step eight, cotton or some other material is placed in the body to give it shape. Step nine, the incision is sewn up. Step ten, tags with appropriate information about the specimen are tied to the legs. This information is likely to include the bird's sex, age, locality, and next-of-kin. Step eleven, the skin is wrapped up and allowed to dry thoroughly. Step twelve, the resulting skin is carefully catalogued, then stored in such a way that it will be a valuable research tool for hundreds of years. For scientific purposes, study skins are perfectly adequate, if not particularly artistic, and a skilled technician with a good caffeine buzz can prepare three or four bird corpses in an hour.

The vast majority of birds at Tring were prepared as study skins, but the Labrador Ducks are taxidermic mounts, as evidenced by their posture and the small holes in the webbing of their feet, which show that nails had held them to whatever display base they used to stand on. Their bases have long since been discarded, and so the birds now lie awkwardly on their sides in a tray in the storage cabinet, as though caught in the grip of rigor mortis.

As with so many Labrador Duck specimens, very little is known about the origin of these birds. The tag tied to the leg of the hen, catalogue number 1863.12.15.27, claims that it was presented to the museum by someone named Verreaux in 1863 and was collected in Labrador. The drake, catalogue number Vel. cat. 42.59a, was presented to the museum by the Hudsons' Bay Company, after being collected in North America somewhere around 1835. With the help of the Hudsons' Bay Company Archives in Winnipeg, Canada, I tried to find out more about the origin of the male by reading the company's correspondence from that era, but all I got was a horrible headache from scanning microfilm copies of 165-year-old letters scribbled by a clerk with awful handwriting.

By all rights, the Natural History Museum collection should have five Labrador Ducks, not two. Another drake and hen had found their way into Rothschild's collection, but they are now housed in the American Museum of Natural History collection in New York (specimens 734023 and 734024) because of Rothschild's romantic indiscretion with the unnamed peeress. Another drake described in a later chapter was offered to the British Museum of Natural History in the late 1940s for £500, but they passed on the offer.

After finishing my work with time to spare before my train, I decided to have a look through Rothschild's birthday present. The museum appears much as it did in the late 1800s. In other cities this sort of presentation has been done very poorly, such that the displays look dated rather than quaint, and the specimens look sad and neglected. At Tring, however, the effect is brilliant. The Victorian cabinets are painted black so as not to detract from the specimens, but with splashes of gold paint as though to remind you that no expense was spared when putting the whole thing together. Elegant wooden doors open to reveal magnificent and exotic insects, and there are gentle reminders to take care in closing the doors because the specimens are fragile. There are stuffed lions and tigers and rhinoceroses and eagles and herons and birds-of-paradise and sturgeon and sharks and . . . well, just about everything! On and on through gallery after gallery. The mind boggles to think that virtually everything in the museum is from Rothschild's own collection. When the mind is simply too full to look at another hummingbird, the museum has a gift shop full of reasonably priced souvenirs, and a coffee shop to recharge the batteries. On its wall I spied a picture of Rothschild in a carriage drawn by a zebra, the perfect image for so eccentric a character.

Cabinets full of dead animals in public museums are not for everyone. But, with proper interpretation by skilled tour guides, this sort of exhibit can go a long way toward helping us appreciate the unity and diversity of the natural world. Some see the cabinets full of eggs and stuffed birds in scientific collections as a waste, but judicious collecting has very little impact on bird populations and has helped ornithology to advance as a scientific endeavor. Whatever may have driven the Labrador Duck to extinction, it certainly wasn't collection for museums, and without those few stuffed specimens, we would have no tangible reminder of the finality of their elimination.

THE COACH TRIP to Cambridge was just about enough to make me rethink my fascination with the British public transport system. Even though my trip was just 80 miles each way as the duck flies, the return journey took seventeen hours, including a forty-five-minute stop to examine Cambridge University's Labrador Duck. Advertisements promoted National Express as "Britain's Coach Network," offering seventeen buses a day between London and Cambridge at a fare of just £8.50. What a deal!

Lisa took a day away from her summer research at a pharmaceutical facility in Kent to see her first Labrador Duck, my sixth. Lisa and I got the last two seats on the coach to Cambridge. We were told we were really lucky to get seats, because most departures before early afternoon were fully occupied by travelers who had paid an extra £3 to ensure their seat. Hmmm—it seems to me that if everyone has to pay an extra fee, then the price isn't really £8.50. Naturally, we couldn't get seats together. Lisa was seated beside a fragrant lady, and I was wedged between two gentlemen from France and a fragrant bathroom, whose door was held shut by my foot jammed against it. Everyone on the bus was either hot, grumpy, and sleepy or hot, grumpy, and asleep.

London is one of the most exciting and vibrant cities in the world. That is a given. However, from the Victoria coach station, it took the express bus forty-five minutes to clear urban sprawl and get up any speed. After two minutes of greenery, the coach slowed down as it entered another endless stretch of shops and homes. On the second-hottest day of the year, such slow progress rubs some of the sheen off London and its boroughs.

After arriving at the Cambridge coach terminal, we ate our picnic lunch in the shade of an adjacent park. My seventy-two-year-old map of the town told us that the green space was called Christ's Piece. Christ's piece of what? We watched visitors getting henna tattoos, and then joined the tourist throngs watching four stern but polite police officers question a young man about a video camera for which he didn't have a receipt. A short walk down Drummer to Emanuel to St. Andrew's to Downing took us to the New Museum Site. The museum site didn't look particularly new, but we were told it was new about one hundred years ago, and they just hadn't gotten around to renaming it.

Ray Symonds, collections manager of Cambridge's University Museum of Zoology, was the first museum curator I had met who looked like what a museum curator should look like. Charming and quiet, perhaps a bit pale, bespectacled, and with a tidy beard, Symonds seemed to be the type who loved animals but was a bit too timid to actually go outside to chase them down. He oversees an impressive collection of 32,500 bird specimens and 10,000 clutches of eggs. I asked Symonds why the museum had gone to so much trouble to make the details of its collections available online but had failed to include information about its specimens of extinct and endangered species. It seems the museum was particularly worried about people with an unhealthy fascination with parrots, such that they might try to stroll off with a stuffed specimen, or travel to the spot where an endangered parrot was collected and try to bag one for themselves.

Labrador Duck 6

Prepared as a study skin, the adult drake in Cambridge is by far the ugliest specimen in the world.

The Cambridge specimen was acquired by a Mr. A. Strickland in 1850, although it is not clear where he got it. Mrs. H. E. Strickland gave the specimen to the museum in 1867. It was my first Labrador Duck prepared as a study skin rather than as a taxidermic mount. An adult drake, it is housed in a plastic bag, presumably to keep all of its bits together. He looks as though he had been hit amidships by a terrible shotgun blast, and that emergency veterinary support was not near at hand to fix him up. His head is floppy and held onto his body only by the stuffing in his neck. The midline incision where his guts were removed has not been stitched up. The protruding stuffing appears to be a combination of plant fibers and animal hair, but I wasn't willing to grab a handful to stick under a microscope; scientific curiosity has its limits. As with most study skins, he doesn't have glass eyes, and bits of stuffing poke out through the orbits. The specimen is a perfectly adequate presentation for the measurements that I wanted to make, it just isn't very artistic. Lisa thought it looked sad.

Curiously, the museum also has the breastbone and wishbone from the body of one of the Labrador Ducks in Liverpool. That duck was shot by J. W. Wedderburn in Halifax, Nova Scotia, in April 1852. It seems that Alfred Newton, Cambridge professor of zoology and comparative anatomy from 1866 to 1907, was convinced that the best way to figure out how birds are related to each other was to examine their breastbones, and so the museum wound up with a hell of a lot of them. These are the only skeletal remains of the Labrador Duck anywhere in the world.

Having finished my examination of the duck, Lisa was feeling woozy. This was probably the result of the heat of the early afternoon and the smell of mothballs used to keep the stuffed specimens free of pests. The British Occupational Health and Safety Board would do well to check the ventilation system in Symonds' work area. We wandered the streets close to the museum, hoping that some food would allow Lisa to feel better. We found a nice café down an alley beside a church, but Lisa couldn't face anything but water.

Looking for shade, we went back to Christ's Piece beside the bus station. We watched a group of young lady tourists from Japan get confused about the intentions of a panhandler, and marveled at the elegant razor wire that kept people in the park from getting into surrounding college residences. Lisa felt much better after she had thrown up in a public toilet near the coach station.

Given our short stay and Lisa's queasy tummy, we didn't get to see much of Cambridge. So what can I tell you about the city? It has a system of about twenty really good university colleges that I wasn't smart enough to get into as an undergraduate student and am not smart enough to teach in as a professor. As a seat of higher learning, Cambridge dates to the twelfth century. Friction between the university and the town led to a riot in 1381, such that several colleges were sacked. The streets between the bus station and the New Museum Site are lined with magnificent buildings and inviting pubs. You can see lots of engaging architecture and lush gardens from the window of the National Coach bus. Beyond that, I am no help to you at all. You'll have to go there yourself. I am sure that the folks at the tourist information center are all very pleasant and helpful.

IMAGINE YOURSELF IN a pub with a few too many drinks under your belt, or wherever you happen to keep your drinks. Your friends might decide to celebrate your inebriation by putting you on a train bound for someplace exotic; Liverpool, for instance. If you were to wake up with a blinding hangover at the Liverpool train station, you would be faced with four problems. The first is your obvious drinking problem. The second is that you have been hanging out with the wrong sort of friends. The third is that your friends, just to make the gag a little funnier, would probably have taken your wallet, leaving you penniless. They may also have taken your trousers. The fourth problem is that you would be hard pressed to figure out where in the hell you were. Liverpool's Lime Street station is just like any of Britain's other train stations of similar size. It is an anonymous, boring, noisy enclosed space. Perhaps that is why all of the people working in the Lime Street station are just a little grumpy. The bathrooms are well hidden, the ticket agents are too few, the coffee is overpriced, and the massive clock on the wall runs exactly one minute fast.

You can solve your fourth problem by looking at the signs proclaiming "Lime Street," as soon as you are able to pry your eyelids apart. The first and third problems I can't help you with. Perhaps some well-placed collect telephone calls would be a good start. As for making new friends, you may be in luck, because just beyond the train station's front doors is a magnificent city whose reputation does not adequately reflect its charms, and it is full of people worthy of your friendship. Please trust me, you really must go to Liverpool right away, and you don't even need to wait for your reprobate friends to help you. Best of all, you will be in a city with no fewer than three stuffed Labrador Ducks.

When you arrive in the great city, turn your back on the train station, and your first sight will be St. George's Hall. St. George's is really astonishingly big. Its scale probably exceeds that of the parliament buildings of most Commonwealth countries. The promenade is guarded by four stone lions. A much-larger-than-life statue of Queen Victoria riding sidesaddle matches a statue of Albert, described on the base as "a wise and good prince," who was at least wise enough not to ride sidesaddle.

If St. George's Hall looks as though it needs a good scrub on the outside, don't despair. Inside you will be greeted by a vaulted ceiling with images of Neptune and Roman soldiers, of angels and cherubs. There are lustrous marble columns, statues of dead mayors and parliamentarians, ten incredible chandeliers, a pipe organ, and a stained-glass window of St. George being mean to an oversized lizard. Behind the building are gardens, walkways, benches, more dead mayors, a dead prime minister, and a tribute to the King's Liverpool Regiment. The whole effect leaves you thinking that, as saints go, George must have been a really, really good one; if not exactly best friends with God, then certainly on a first-name basis. Oddly enough, St. George was an Arab who died in Palestine around AD 303. He was adopted as England's patron saint by the crusaders of Richard the Lionheart eight hundred years later.

Next to St. George's Hall is a column dedicated to Wellington, the County Sessions House, Walker Art Gallery, International Library, Central Library, and Liverpool Museum. This would be a good time to take a breather and sit on the steps of the museum. Within a few minutes, you will get an eerie sensation as your bum starts to vibrate. This is nothing more erotic than the rumble of the underground train passing beneath the building; briefly titillating and well worth the wait.

Lisa was in Liverpool to attend a Physiological Society conference. The theme was something to do with the amount of calcium inside the body's cells. At social gatherings, I was swamped with talk about uploaded G-protein dojiggers and the cascade system of frizzle-bibble membrane incorporation. After a while, it all sounded like "bzzzz, phzzzz, bzzzz" to me, and the strangest part is that physiologists continue to speak that way even when they are drinking. They claim that this kind of research is terribly important and will probably save me from dying of a horrible wasting disease. In contrast, I was in town to see the Labrador Ducks held by the National Museums and Galleries on Merseyside.

By now you are familiar with my routine for museum visits. I contacted the person in charge several months in advance, backed this up a month or so before the big day, finishing with one final message with about a week to go. On the whole this process seemed to work pretty well. In the case of Liverpool, the person in charge was Dr. Clemency Fisher, Curator of Birds and Mammals, and we traded all of the necessary email messages. Lisa and I booked a room at the University of Liverpool, and the train tickets were paid for. What could possibly go wrong?

Just a few days before Lisa and I departed for Liverpool, Fisher sent me a message: "Glen, really sorry about this, but the collections are closed to visitors at the moment (decision taken by the Head of the Liverpool Museum). Hope you get this in time!" I was gutted. The plan was to see *all* of the Labrador Ducks in the world, not all except these three. For an overseas researcher on a limited budget, this was pretty much a one-shot thing; it was now or never for me and these ducks. I wrote back to Fisher, pleading, groveling, and begging for the head to reconsider her decision.

Fisher clearly felt embarrassed that a poor administrative decision made by a poor administrator meant that a legitimate researcher was going to be denied access to a collection intended for research. I suspect that embarrassing Fisher is a big mistake. She mobilized her colleagues to work on the problem. The museum's head of science came up with a useful suggestion—since there was a ban on visitors to the collection, but not on loans of specimens to other institutions, it should be possible to loan the Labrador Ducks to another museum without breaking the new regulations. Indeed, the receiving institution could even be inside the same building, as long as I didn't actually visit the collection. And that is how I came to examine the Liverpool Museum's Labrador Ducks in the Natural History Centre, up a floor and down a hall from the museum's actual collection rooms. For all of Fisher's efforts, and those of her colleagues, particularly against the ravages of an overblown administrator, I am truly grateful. By comparison, St. George's battle against the giant gecko was probably a cakewalk.

The Labrador Ducks exited the research collection of the Liverpool Museum at 9:30 in the morning, and reentered the collection a little before 13:00. They were officially on loan to the Liverpool Museum's Natural History Centre. The reason given on the exit receipt was "Research Loan." Given the value of these specimens, it is probably the last time they will ever make a trip outside of the collection. I felt that I was creating an awful lot of bother.

Fisher was a star about it all. She is an imposing figure of a woman who immediately makes you feel that you are the most amazing person in her life. Her father was also a noted ornithologist, and I have been told by an unreliable authority that her grandfather was Geoffrey Francis Fisher, the Archbishop of Canterbury for all of the big events early in the reign of Queen Elizabeth II, including her wedding. Perhaps Fisher got her welcoming warmth from that side of her family. On the day of my visit to the museum, she had just had her hair dyed purple to match her scarf. I really appreciate that sort of free spirit; perhaps that approach to life came from her mother's side of the tree.

Over lunch in the Walker Art Gallery's coffee shop, Fisher filled me in on exactly why the collection had been closed to visiting scholars. During renovations to the museum, and moving specimens from their regular housing, into storage, and then back again, it became apparent that some of their bird specimens had gone missing. It is hard to imagine why some of the missing specimens would have been taken, being as common as dirt, and not particularly well known for their decorative qualities. However, some very valuable specimens of extinct species had also strolled off. In a collection of that size, it isn't impossible that the missing specimens were simply misplaced in some dark corner. In any case, the situation called for a forensic audit, and a review of the protocol for visiting scholars. Fair enough, but the head of the Liverpool Museum apparently used my case to flex some muscle, despite my offer to submit to a strip search after my visit.

How did Liverpool come to have three Labrador Ducks in its

collection? In 1485, Sir Thomas Stanley was made the first Earl of Derby by King Henry VII, in gratitude for being rescued at the Battle of Bosworth Field. From then on, the Stanley family always seems to have been on the winning side of each armed conflict. That, and some mining interests, left the family very well off. Liverpool was decimated by the plague in 1548 and again in 1558, but the Stanley family continued on, sequestered in the Tower of Liverpool from rats, their fleas, and the bacteria they carried. Fast-forward a few earls to October 1834, and the death of the twelfth Earl of Derby. This generous act made his son, Lord Edward Smith Stanley, the thirteenth Earl of Derby. A bit like Walter Rothschild, Earl 13 had a passion for collecting, as well as the money to do something about it.

The year before his father's death, Earl 13 purchased two Labrador Ducks from John Gould: an adult male (D920) and an immature male (D920b). It was still early days for Gould, who later became renowned as an artist, a businessman, and a great bird enthusiast. Earl 13 apparently thought that the drably colored bird he got from Gould was a female, although it appears to me to be a very young male. Earl 13 was also given a female Labrador Duck (D920a) by Thomas C. Eyton, a Shropshire magistrate, on February 18, 1840. With the death of Earl 13 in 1851, and the ascendancy of Earl 14, all three Labrador Ducks and 15,000 other specimens passed by bequest to the Liverpool City Council, which served as the foundation for the Liverpool Museum's natural history collection. Today the museum has about 55,000 bird specimens.

The next specimen (T9597), an adult male, was shot by Colonel J. W. Wedderburn in Halifax Harbour, Nova Scotia, in April 1852. It was given to Canon H. B. Tristram in November 1876 and received by the Liverpool Museum in 1896. As mentioned earlier, the breastbone of this male was put into the collection of the University Museum of Zoology in Cambridge in 1879.

By my count, that makes four Labrador Ducks, and yet the Liverpool Museum has only three. One of the adult males had gone missing, and the curatorial staff wasn't clear when or how it disappeared. The missing Labrador Duck wasn't the only bird specimen to go astray at some point in the distant past. In 1959, R. Wagstaff, Keeper of the Liverpool Museum's Department of Vertebrate Zoology, responded to Paul Hahn's questionnaire about stuffed extinct birds. He wrote to say that the museum had eleven Passenger Pigeons, an Eskimo Curlew, three Carolina Parakeets, and three Labrador Ducks. Wagstaff also explained that three stuffed Whooping Cranes were listed in the museum's catalogue but could not be found. How do you misplace something as big as a Whooping Crane, and something as cute as a Labrador Duck?

The answer is quite simple in the end. After my examination of the ducks, Steven Cross of the museum's Nature Centre gave me a tour of the facility. Along with many amazing natural history artifacts on display, he showed me striking before-and-after pictures of the museum, in this case before-and-after bombing during World War II. Early in the war, most of the collection had been removed to a system of caves in Wales, just in case of that sort of attack, but many specimens were left on public display, apparently including a Labrador Duck and three Whooping Cranes. It would be difficult to prove, but these birds appear to have been casualties of war. If they weren't blown to smithereens, they were likely carried away from the rubble by children the morning after. Kids are like that. There are probably still a few such artifacts in the attics of Liverpool homes.

I don't mean to minimize the devastation of war by making the loss of a stuffed duck sound like something important. Liverpool was Britain's major port in World War II, and the Germans made a concentrated bombing effort on the docks, particularly on three nights in May 1941. Bombs don't always fall precisely on their targets and the air raids resulted in the death of more than 2,500 people, and serious injury to a similar number. More than half of Liverpool's homes were damaged and 11,000 were totally destroyed. As a tribute to the human spirit over bombs, cargo was being handled again at Liverpool within a week, and the docks were operating normally within a month.

Labrador Ducks 7, 8, and 9

So here is what I found in examining the three remaining Liverpool ducks. The adult drake, the one not bombed by Germans, has only one glass eye, and his left foot looks as though it has been nibbled by mice. The taxidermist may have made him a little plump, but otherwise he is in good shape. The second bird has been described in the past as a female, but subtle markings on the bill, and lighter feathers on the breast, head, and neck suggest that it is probably a very young male. His tail and wing feathers are a bit beaten up, but he is otherwise getting by. The third individual, presumably a female, isn't doing quite so well. Both legs are broken, and her right foot isn't attached to the rest of her body. Instead of standing, she now sits forlornly on her belly on a small wooden base. When the time comes, some creative taxidermist will probably have to rebuild her, using the legs from another, less valuable, duck specimen. She is mottled brown and gray and white, as befits a hen that would have been trying to hide from predators as she sat on her eggs. These three remaining specimens are kept together in a well-crafted wooden box in a large cabinet along with other valuable specimens. A small card on the box explains that Wagstaff examined it on February 15, 1962, and found that all of the ducks were in place, and that someone else tossed in a Vapona no-pest strip on March 31, 1980, to treat a bug infestation. Museum curators really, really hate bug infestations.

I accomplished all of that in my first full day in Liverpool, leaving me with two full days to slob around and see the sights while Lisa listened to talks about inward rectum calcium channels and ATPase-activated smooth-muscle death rays. I toured museums, and helped visitors from France get hopelessly lost. I helped some German tourists by taking their photographs in front of the Mersey. I discovered that the "Liverbird," an ugly cross between a cormorant and an eagle, was added to the city's coat of arms in 1797. I took in an amazing exhibition on transatlantic slavery, which was one of the major factors that allowed Liverpool to become such an important city.

On my last day in Liverpool, I checked my email and found about twenty messages. Most of them were work-related and so I was able to ignore them. But a message from one's mother is one that can be ignored only at the risk of a few millennia in purgatory. In her message, my mom caught me up on family news from Canada. She also said that the Chilton family had an interesting link to my current adventure. It was from the once mighty docks of Liverpool that my parents and older brother had departed England late in April 1954, looking for a better life in Canada. They had sailed from Liverpool's docks on RMS *Ascania* on her last transatlantic voyage.

I took the long downhill walk from the University of Liverpool to the dock area. Half a century ago, the area would have been a far more serious place, involved in commerce and insurance as it had been for a couple of centuries. Today, ships continued to sail in and out of Liverpool harbor but at a much reduced rate, and much of the waterfront has been reborn as upscale shops, museums, and eateries. There was no way for me to know which particular bit of the Mersey my family had departed from, so I spent an hour walking up and down the waterfront. My parents had taken my genes out of Britain through Liverpool in 1954, and now, a half century later, I had brought them back, even if only for a while. My time in Britain was done.

Chapter Five

A Swelling in My Socks

Even though I come from a country that is officially bilingual, I can claim to be absolutely horrid with foreign languages. My grasp of French is limited to a handful of words like *jambon, crayon, chien*, and *bibliothèque*. These disconnected words are unlikely to be of any use unless I find myself in need of a pencil in a library to make notes about types of ham preferred by dogs. And so it was that with three Labrador Duck adventures awaiting me in the far-flung corners of France, I knew it was best to engage a minder. I needed someone who could ensure that I got a hotel room and not a room in a brothel, and who could assure me of a glass of wine and not a prison record. I needed someone to keep me calm by booking train tickets during an impending general strike, and to translate technical expressions like *duck* in natural history museums. This special someone would be my guide to some of the most beautiful cities in Europe. In short, I needed Julie Rainard, a biology student just returned to Paris after completing undergraduate studies in England, and a work colleague of Lisa's.

For all of its glamour and efficiency, the Eurostar train deposited me in Paris's Tenth Arrondissement, often described as a conglomeration of commercial zones and sex shops and best avoided. I was met at Gare du Nord by Julie, holding a placard with my name on it. Julie is that most precious of all commodities—a beautiful woman who doesn't seem to know that she is beautiful. She has large and sparkly eyes capped by razor-thin eyebrows. The right eyebrow has a narrow break, the result of an endearing unconscious nervous habit of touching it repeatedly in one place. Her pouty lips are just the sort that keeps Julia Roberts salivating with envy. Her hair is the color of Cadbury's milk chocolate.

I have an amazing trick for ensuring that Lisa doesn't get jealous when I am on the road with beautiful young women like Julie. It probably belongs in the *Handbook for a Happy Relationship,* wedged somewhere between "Spend at least as much time listening as talking," and "Don't murder your in-laws." My little trick is this: "Never, under any circumstance, give your partner a reason to doubt your fidelity." That's it. No need to thank me for the advice; that's what I'm here for. In twenty years of marriage, I have never given Lisa any cause to suspect that I was doing something that I shouldn't, and so she trusts me implicitly. After all, there is nothing so powerful as the company of another woman to remind me how fortunate I am to be with Lisa.

In the weeks leading up to my trip to France, it was clear that engaging Julie had been exactly the right thing to do. Her efficiency and persistence with ticket agents and museum administrators showed, but the first hour after my arrival demonstrated just how perfect the decision had been. She met me with a Métro ticket in hand. She had booked a great inexpensive hotel room, and had ridden the Métro to my hotel the day before my arrival so that we wouldn't have to waste time searching for it when I arrived. In short, my adventure was going to go as smoothly as adventures ever do.

OUR FIRST FRENCH Labrador Duck adventure was, conveniently enough, in Paris. However, the museum wouldn't open until Monday morning, leaving Julie a whole day to give me a taste of her fair city on the Seine. We started with a tramp up Montmartre hill to the Basilique du Sacré-Cœur, which should probably be famous for housing a portion of the Sacred Heart of Christ, but is probably more familiar from a scene in the film *Amélie,* involving a telescope and arrows drawn on the pavement with flour. The view of Paris from Sacré-Cœur is unbeatable, but those travelers who are committed to the very, very old should look elsewhere, as the basilica was not consecrated until 1919.

The Montmarte region around the basilica had been a hotbed of artistic energy until tourists found it. Even now it is a very fashionable address occupied by many famous Parisians I had never heard of. After passing an older lady in sunglasses walking two small dogs, Julie pointed out that she had been quite a famous actress in her day, although she could not remember the lady's name or any of her films.

We had lunch in a restaurant district south of the Seine, offering every form of culinary delicacy I had heard of and a few that I hadn't. We then visited the great Cathédrale Notre-Dame de Paris on the Île de la Cité, the largest island in the Seine. Rather oddly, this cathedral devoted to Our Lady is only one of ten by that name in France. This particular Cathédrale Notre-Dame has a long and glorious history as a living church. Construction began in 1163 on the site of a Roman temple, and it narrowly survived the French Revolution. It probably represents France's second most frequently photographed monument after the Eiffel Tower.

No one would argue that Notre-Dame isn't a really whiz-bang church. The architectural superlatives go on and on, as do the really keen stories. For instance, the cathedral can house 9,000 worshippers, and is 425 feet long and 115 feet high. On the facade, above the portals, is a parade of statues of the kings of Israel and Judea, which were apparently pulled down in 1793 by people who thought they represented the hated French kings, but later restored to their pedestals. The cathedral houses such relics as a fragment of the True Cross, a bit of the Crown of Thorns, and a Holy Nail. Unlike at Sacré-Cœur, there was no mention of Holy Body Parts. Despite all of this, it is not the sort of place that I wanted to stop for a good long chinwag with God. The stained-glass windows beggar description, but they leave the interior rather gloomy. With endless hordes of visitors, quiet contemplation was right out of the question. The gift shop inside competed with hawkers outside the church. This magnificent monument felt a little less like a house of worship, and a lot more like a very elaborate museum.

Past the pyramids of the Louvre and through the Place de la Concorde to the Champs-Elysées, Julie and I wove our way through grand gardens perforated with opulent fountains, gawked at a 3,200-year-old Egyptian obelisk brought to Paris from Luxor, and passed innumerable gold statues of heroes with and without horses. The city was a riot of people enjoying the sun and warmth of the Sunday in the company of friends and strangers. We saw the start of a 2,000-strong in-line skating cavalcade in support of the fight against AIDS, and then picked up the procession as it doubled back along the avenue.

At the western end of the Champs-Elysées is the Place Charles de Gaulle. Twelve major streets converge on a giant traffic circle at the Place, and sitting squarely in the middle is the magnificent Arc de Triomphe, one of only four arcs de triomphe in France. Guidebooks rabbit on and on about the history and dimensions of the Arc, but what most don't tell you is that you can climb to the summit for an astonishing view of the city. The top observation platform at the Eiffel Tower may be a tad higher, but it also has great snaking lineups. We walked through a tunnel to the center of the Place, and, without having to wait in any lineup, handed over a few euros. By the time we finished with the 296 stairs to the top of the Arc, I was a little dizzy from the heat and oxygen starvation, but that did not diminish the experience at all. Julie pointed out great structures that constituted the Parisian skyline, but also the near absence of construction cranes, something that she considered to be a blight on the landscape of London. With the childlike glee of an ornithologist I looked down on swifts swooping for insects.

As we gazed on the Place Charles de Gaulle, the driver of a small, cheap car lost his confidence and brought it to a screeching halt in the giant traffic circle. Eight other cars became trapped behind the first car, as everyone else rocketed around them. Ten minutes later, all nine ensnared cars were still there. According to Julie, in 1986, a Volkswagen Beetle stalled going around the traffic circle, and the driver could not restart it. In the next forty-five minutes, more than seven thousand cars became hopelessly enmeshed in the Place and adjacent streets. Since no one could decide how to deal with the problem, the cars were abandoned, and travel in western Paris came to a virtual standstill for seven months.

• • •

BUT, OF COURSE, in this city of splendors unending, my voyage was all about the splendors of Labrador Ducks, and so bright and shiny early on Monday morning, Julie and I were off. We followed the directions provided by my contact, Dr. Claire Voisin, researcher at the Laboratoire de Zoologie, Mammifères et Oiseaux, Muséum national d'Histoire naturelle, in the Fifth Arrondissement. Down rue Buffon, past old unnumbered buildings, through a gate and across a yard to an arch with four doors, to door number three. Push a button, climb a wooden staircase, and there we were—welcomed by Voisin into the behind-the-scenes world of the museum's ornithological research collection. I asked Voisin how the collection had avoided destruction in World War II when so many other museums in Europe had been leveled by bombs. Voisin indicated that salvation had been sheer luck; a building just down the street had been demolished in just that way.

Labrador Duck 10

The stuffed adult Labrador Duck drake is one of many great ornithological treasures in the care of the Muséum national d'Histoire naturelle. According to documents from 1935, the Paris duck was donated to the museum by MM. Milbert and Hyde de Neuville in 1810. Jacques-Gérard Milbert was born in Paris in 1766 and died there in 1840, spending much of the intervening period as a professor of drawing in the great French capital. As a break from this vocation, which probably involved sketching a lot of nude models, he traveled to the United States as a naturalist in 1815, returning to France in 1825 when he ran out of cash. His voyage to America was in the company of Hyde de Neuville, the French consul-general at New York, who went on to facilitate Milbert's extensive travels in the American East and South. If his voyages took him north into Canada, he might have collected the Labrador Duck on the breeding grounds himself. Perhaps he collected it on the wintering grounds of the northeastern United States. Otherwise he and/or Hyde de Neuville probably traded for it or purchased it—the records offer no hint of the real story. Yet we do know that Milbert contributed many specimens to the Paris collection.

The Labrador Duck was housed in an uninspiring gray metal cabinet along with two Pink-headed Ducks, four Great Auk eggs, and an assortment of other extinct and precious items. Julie waited patiently as I slowly pored over the drake, poking and peering and measuring with calipers and dividers. Trying, but failing, to improve on nature, someone in the specimen's murky past had painted portions of the bill mustard yellow. The most peculiar feature of the Parisian Labrador Duck is that its feet are not its own. The base on which it stands is careful to point out that "*les pattes sont factices*," "the feet are fake." The duck's original feet were nibbled by mice and replaced by those of a Mallard. At some point the specimen's eight-sided wooden base had been painted gray-white, but the painter had been too lazy to remove the duck first, and had slopped some paint on the feet. I filled Julie in on what I was doing as I proceeded to examine my tenth specimen, and in a little less than an hour we were done.

As Voisin led us toward the exit, we passed a number of large white cabinets labeled TYPE. I guessed that the word meant the same thing to a French biologist as to an English one, but I asked anyway. Yes, indeed, said Voisin, the cabinets were filled with valuable type specimens; the museum owned a couple of hundred of them. These are the individual specimens that taxonomists use for comparative purposes to represent the whole species. For perspective, you might keep in mind that Canada's National Museum in Ottawa does not have even a single type specimen of a species of bird. She gladly pulled out a number of cormorant and penguin type specimens. Most had been prepared as taxidermic mounts, but had been removed from their wooden bases and restuffed as study skins, lying flat on their backs in plastic storage boxes, in order to take up less space and to minimize the risk of damage.

JULIE SET OFF for home and I set off in search of dinner. Being an exceptional hostess, Julie had provided me with a very thoughtful gift—a French phrase book to lead me through those awkward little social encounters when she wasn't available to keep me out of trouble. Ordering dinner, for instance. The book provided me with helpful suggestions for conversation starters, including "*Pouvez-vous me donner des bon marché préservatifs, s'il vous plaît?*" (Can you give me some inexpensive condoms, please?). The book suggested that I try such helpful French expressions as *"Permettez-moi de vous offrir quelque chose à boire."* (Let me buy you a drink, baby!), followed by *"Si on allait dans un endroit un peu plus calme?"* (Shall we get out of this joint?). If the first request fails, but the second and third work out, the guide suggests that I might be listening to a physician say, *"Vous avez une inflammation de les chaussettes"* (you have an inflammation of your socks), to which I am apparently supposed to respond, *"Je voudrais du citron et une couche!"* (I would like some lemons and a diaper!). This sort of dialogue left me wondering how much Julie had paid for the phrase book.

Even so, I became quite dedicated to the little book. It did, however, overlook the one key phrase that might be more valuable than any other. It would go something along the lines of "Pardon me, but I am astonishingly useless at French. Instead of having me mess up your beautiful language, if I point at something, will you please package it up in exchange for a handful of euros?" In the absence of that altogether invaluable phrase, the book still allowed me sufficient hacking and slashing of the French language to obtain milk, a vegetable panini, and raisin buns from an assortment of shops.

I took my culinary treats to a park close to the police station and sat on a bench with all of the other patrons enjoying a quiet moment in the early evening's warmth, after a long day of work. I ate while watching children play in the shade under the careful eye of their keepers. A couple of pigeons wandered by, hoping for a handout. Julie had warned me that Parisians are not all so fond of pigeons as I am, and so I left the little fellows to forage for themselves. My reserve held until a particularly pathetic little pigeon hopped by on one leg, trying to hold the stump of his amputated second leg out of the dust. I ripped a few crumbs from my raisin bun and tossed them to the little guy. An instant later he and I were surrounded by five of his avian friends, followed moments later by five young screaming boys with toy guns and water pistols, trying to shoot the pigeons. The tranquillity of that corner of the park had been shattered, and I was under the full glare of every park patron. I bundled up the remainder of my dinner and fled to find another park.

• • •

Travelers are now able to catch a very fast train between Paris and Strasbourg. Arriving in France one year too early, Julie and I had to settle for a rather more pedestrian but perfectly pleasant journey of about four hours. Slower than an airplane, certainly, but far less claustrophobic, and without the frustrating argument with security personnel about the terrorist potential of a nail file, or the endless wait in a departure lounge, which has precious little to do with either lounging or departing. Not that I would try to judge a region from the window of a train, but the tops of clouds have never taught me anything about anywhere.

I watched as inner city Paris was replaced by sprawling suburbs, then an assortment of agricultural fields, grains, and broad-leafed crops, vineyards, and finally deciduous and coniferous forests. At the journey's halfway point, the vegetation gave the impression that the region was warm and dry, although Julie assured me the area was notably cool and damp. Odd place to grow grapes, then, I would have thought. More and more vineyards spread out on either side of the rail line. We saw a church spire encased in scaffolding, and a giant inflated Ronald McDonald. We whizzed by the Regional Center for the Distribution of Sugar, and a number of small towns, cemeteries, and factories, some covered with graffiti, no less sensible in French than in English.

That we were on a train to Strasbourg at all was somewhat surprising. It had taken seven years of inquiries to get a response about the Labrador Duck in that city. I knew that the duck, a hen, existed, because Paul Hahn's book told me so. He reported that it was from the Verreaux collection and had been taken in Labrador in 1865. Hahn didn't say what or who Verreaux is or was. My letters of inquiry dating back to 1995 had all gone unanswered. But when Julie got on the trail, the wheels immediately started turning. Julie made contact with Dr. Marie-Dominique Wandhammer, Conservatrice du Musée Zoologique, who proved extremely cooperative, sending me fantastic print photographs of the duck and offering to welcome me whenever I wished to see their duck.

With the whole evening free, Julie and I made our attack on Strasbourg. The city is the seventh-largest in France and is the birthplace of pâté de fois gras. It is also the site where the French national anthem, "La Marseillaise," was composed, in 1792. The University of Strasbourg counts Goethe, Napoléon, and Pasteur among its alumni. The city hosts the Council of Europe and is home to a branch of the European Parliament and the European Court of Human Rights. Gutenberg perfected the printing press while in exile in Strasbourg in the fifteenth century. All was incredibly noble and grand.

We crossed a few of the twenty or so bridges spanning the river Ill, a tributary of the Rhine. The best way to see the heart of Strasbourg is on foot, because automobile traffic is restricted in the core. If you go there, do not rely on street names, as they serve only to confuse. Rue Gutenberg is magically transformed into rue des Hallebardes, which goes on to become rue des Juifs, which continues along as rue du Parchemin, which then becomes rue des Récollets before crossing the Pont de la Poste and resuming as rue J. Massol, which finally becomes rue du Genéral Gouraud. You could walk from one end to the other in about ten minutes, and no one living on that street has ever been known to get the correct mail. There is so much wonderful architecture along this street, as along others in the medieval city's core, that Strasbourg has been designated a World Heritage Site by UNESCO.

We treated ourselves to a walk through the eleventh-century Cathédrale Notre-Dame de Strasbourg; tall, ornate, cinnamon brown with dark chocolate sprinkles, and completely awe-inspiring. The architecture makes you want to use words like *garret* and *crenellated* and *Wilhelmian,* even if you aren't sure what they mean. I particularly recommend the sculptured tribute to the Wise and Foolish Virgins, which I completely missed. The cathedral's spire, which can be seen from all over the city, was the tallest one in all of Christendom until some naughty, vain church beat it out in the 1800s. Even so, it makes a great landmark while strolling the city's twisting, turning streets.

To me, Strasbourg has a peculiar German flavor, which is probably not surprising given the number of times it has flipped between German and French rule. Imagine the confusion caused by a street that started off as avenue Napoléon, became Kaiser-Wilhelmstrasse in 1871, switched to boulevard de la République in 1918, was renamed Adolf-Hitler-Strasse in 1940, before finally settling down as avenue du Général-de-Gaulle in 1945. I suppose a local would say that Strasbourg has a distinctive character that is neither fully French nor German. The current dividing line between the nations, the Rhine River, is just a couple of miles to the east.

The following day we set off to the Musée Zoologique de l'Université Louis Pasteur et de la Ville de Strasbourg for my second French duck. At the reception desk, after the usual little song and dance about who we were and why we were there, we found all of the museum staff to be the most incredibly cooperative and happy people imaginable. Marie-Dominique Wandhammer was not immediately available, but we were left in the competent and enthusiastic care of Dominique Nitka, who took us directly to the Labrador Duck. Given the value of the duck, I was astonished to find it sitting on the sort of metal shelving I use to store power tools in my garage. This housing is not particularly dustproof, nor, would I think, theft resistant. In order to save space, the shelves can be run together, but any bird with a particularly long tail is likely to have it rammed into the head of a specimen on the next shelf. The Labrador Duck is simply shoved in with all of the other ducks and geese in the museum's collection.

Labrador Duck 11

Other than being a little grimy, and perhaps a little plump, the hen in Strasbourg is in fantastic shape. Her breast feathers were a mosaic of light brown and gray, and her tail feathers were a bit frayed but otherwise undamaged by her long incarceration at the museum. Someone had gone to a lot of trouble to mount her properly, and the skin around the eyes and the webs of the feet were particularly well done. Attached to the plain wooden base was a small, old, red card that read *Camptoloemus labradorius (Gm.) ♀ Labrador 1865 Verreaux*. I asked Nitka about the collector Verreaux. He dug through the records for a few minutes and found that it was actually the name of a shop in Paris dealing with natural history artifacts. Nitka also provided me with a catalogue of all the ducks, swans, and geese in the collection of the museum. The catalogue showed that fifteen of these birds were purchased from Verreaux dating back to 1856; most of them, like the Labrador Duck, were acquired in 1865. Perhaps Verreaux was having a going-out-of-business sale, and the museum in Strasbourg cleaned up on bargains.

When Wandhammer caught up with us, resplendent in glasses with electric-red rims, she offered us a tour of the museum and asked if there was anything else we particularly wanted to see. I asked if they had a stuffed Great Auk and Nitka replied with pride that they did indeed. The museum had an impressive display dedicated to extinct and endangered species, including the Great Auk, a couple of Passenger Pigeons, and a Carolina Parakeet, behind a sliding glass door that ran from floor to ceiling. When I said that it was a shame that I couldn't photograph the auk through the display's glass door, Nitka dashed off in search of the key, despite my protests about its being too much trouble. When he returned, we found that the glass door opened only halfway, and the Great Auk remained trapped inside. Not put off for a moment, Nitka climbed into the display cabinet to grab it. My first fear was that he was going to drop the auk on some other extinct creature, and then that he was going to break the bird off at the legs as he carried it to a back room for me to examine.

Everyone stepped back to let me have a good look. After all that bother, there was nothing else for me to do but give the poor beggar a good going-over. I stroked my chin and peered intently. I stuck the tip of the small finger of my right hand in my mouth, made "hmmmm" noises, and nodded at the bird. I pointed to spots on the bird as though I was trying to make an important point to an imaginary colleague. The Great Auk in Strasbourg has seen better days, and should probably be downgraded to a Reasonably Good Auk. Moths have taken away a lot of its feathers, giving it an air of mange. It was mounted with its mouth open and I almost thought I could hear it screaming. With most of the feathers around its eyes missing, it certainly had a look of terror. After ten minutes of scholarly peering, I picked the little devil up to return him to his cabinet. Then, thinking better of it, I turned him over to Nitka. If he dropped the auk, he would be reprimanded. If I dropped it, I would be trying to talk my way out of a Strasbourg jail.

THURSDAY WAS ANOTHER day to explore Paris before my final duck adventure in France. Having arranged to meet Julie at 11:00, I used the early-morning hours to explore the district near my hotel. The community was chockablock with six-story apartment blocks, shops selling spectacles or lingerie, and wave after wave of cafés, bistros, and brasseries for casual dining. One of the most impressive features of the Twentieth Arrondissement is the Cimetière du Père-Lachaise. So grand is this 110-acre cemetery that guidebooks give it as many stars as the national museum dedicated to the works and life of the sculptor Rodin. Plaques at each entrance show where the traveler can find the final resting spots of such great persons as Ney and Masséna, Abélard and Hélöise, Piaf and Toklas, and a host of other celebrities that I had never heard of. The composer Chopin is buried at Père-Lachaise, but his heart apparently resides in Warsaw; no explanation was offered for the dissection. One of the most infamous residents of Père-Lachaise is rock-and-roll legend Jim Morrison. I have been told that stoned visitors to Morrison's grave site cause such disruption that the city looked into the possibility of having him disinterred and removed to another site, but found that they had no legal right to do so. And so the Lizard King sleeps on in the City of Love.

To me, the most impressive feature of the cemetery was not the personalities that rest there, but the opulence of all of the other grave sites, unlike anything that I had ever seen. The lives of some people have been commemorated by giant slabs of marble or granite. Some of these are turned on end to create great *Space Odyssey*–like monoliths. Other persons rest under enormous statues, some of half-carved individuals striving to leave behind their earthly shackles and ascend to heaven. Some statues are carved with a series of skulls and demons and a single angel. A few monuments have been embellished with large copper medallions, which have been oxidized by time and rain to leave long hideous green stains.

In this cemetery, many persons, perhaps most of them, are remembered with mausoleums about the size of a queen-sized bed but rising 16 feet or more, although some take on the proportions of a small chapel. Many of these structures recognize a family or two rather than a single individual. The mausoleums have locked twin doors of iron, many of which are rusting away to nothing. A fair few have stained-glass windows featuring images of Christ, the Virgin, or both. Many windows in older tombs have been broken by vandals or by time.

• • •

Being reasonably dedicated to the phrase book Julie had given me, I was very proud when I could get out "*Demain, je voudrais le petit déjeuner, s'il vous plaît*" without inciting scornful looks. If pressed, I could even convey I was residing in room *cent trois*, and I would like my breakfast at *huit heures*. Even so, I generally relied on five key expressions. These were: *bonjour* or *bonsoir, s'il vous plaît*, numbers between *un* and *vingt, merci*, and *au revoir*. Much could be accomplished with those words and a bit of pointing.

But I certainly had my limitations. A disturbing event occurred when Julie and I were on the Métro, but it came without any announcement at all. Rocketing along in a dark tunnel between stations, the driver suddenly threw the brakes into an agonizing squeal, leaving the passengers to catch one another. I had a good grip on a rail, and was able to throw out my left arm to catch a lady before she crashed into the front of the coach. She thanked me four times. We were then left in darkness, save for the emergency lights. The train had stopped dead. Eyes darted, and a few people giggled nervously as a minute or two passed. Being in the front car, we could hear the shouts of at least three voices. They came at me too fast and I couldn't make sense of a single word. The train started again, and we detrained at the next stop. Hoping that Julie could fill me in, I tried, "A rather quick stop!" as an opening line. It probably hadn't occurred to her that I was the only person on the car who had absolutely no idea what was going on. She explained that someone had tried to kill themself by jumping in front of the train in the darkened tunnel. All credit to the driver, who spotted the person in time and stopped just shy of disaster. I was absolutely shattered by the news. I was having one of the greatest days of my life in one of the greatest cities in the world. And yet just a few yards away was a person so tortured that a dramatic and violent death seemed to be the only solution.

I think I can claim to have learned a few small things about language, transportation, and general behavior while in Paris. I thought of them as my ten rules for getting by and staying alive in the French capital:

1. Your place in a queue is only a state of mind unless you are willing to defend that place with tooth and nail.
2. The penultimate accessory to disguise oneself as a Parisian is a baguette in a paper bag. The disguise can be improved upon only with a cigarette.
3. Do not be deceived by the number of letters in a French word; most words are pronounced with only one syllable, if that. Never pronounce the second half of any word.
4. Do not be deceived by the number of words in a French utterance. The sentence "*Mon à nôtre la fenêtre votre carotine d'agréable depuis en petite dix-neuf avec Caroline et Antoine . . .*" probably translates as "The train is five minutes late."
5. While on foot, do not be deceived by what appear to be pedestrian crosswalks. They were installed some years ago as a joke, and are now used as an opportunity for target practice by drivers. And lawyers.
6. As a driver, do not be deceived by an apparent right-of-way over pedestrians. In Paris, those on foot will cross the road anywhere and at any time. The bravest pedestrians are the most elderly and otherwise least mobile.
7. Try not to require an ambulance on a Friday afternoon. Flashing lights and a siren give an emergency vehicle no priority at a traffic circle between 16:00 and 19:00. In case of life-threatening injury, consider walking to a hospital.
8. Do not be tempted to rent a car in Paris. The last vacant parking spot was reported in 1987, and four men died in the battle for it.
9. In Paris, street vendors without licenses are afraid of the police. Beggars are not afraid of God Almighty.
10. In Paris, beggars are fluent in every language ever devised. They should be employed by the United Nations as translators.

ANYONE WHO FINDS himself in France, but doesn't take the opportunity to visit the northern city of Amiens, should give himself an enthusiastic kick in the backside. Sitting astride the river Somme, dissected by narrow streets and a system of canals, praised for the fertility of its fields by Julius Caesar, the final resting spot of visionary author Jules Verne, and home to a really tip-top cathedral, no one should miss it.

On Friday, Julie and I found ourselves on an early-morning train north out of Paris. Our compartment was sparsely occupied, mainly by salespeople in cheap business clothes, with the sort of look in their eyes that said: "Gotta make a sale in Amiens . . . gotta make a sale in Amiens!" We were headed for a 10:00 appointment with the director of the Musée de Picardie, Monsieur Matthieu Pinette, who oversees the operation of a large and really top-shelf collection of art and archaeology. To the surprise of no one, we were not after insight into the world of fine art but rather an elusive duck.

In 1897, Mr. J. H. Gurney of Keswick Hall, Norwich, England, published a very brief note in the ornithological journal *The Auk*. Gurney wrote: "In the Museum at Amiens in France, which is located in a temporary and very unworthy building by the river, I was surprised to come across a fine adult male Labrador Duck, *Camptolaimus labradorius*, in good preservation." He went on to speculate that it may have been sent to Europe sometime before 1850 by John Akhurts. Hahn, in his 1963 book about stuffed Labrador Ducks and other extinct North American birds, repeated these few details. He seemed quite certain there was a Labrador Duck in Amiens.

My pursuit of this duck specimen began in 1995. It doesn't sound like much of a challenge, right? You simply write to a museum and ask about their duck. It didn't help that I was unable to find any contemporary reference to a natural history museum in Amiens. Over the following four years, I wrote letters to anyone I could think of, including the local zoo and a brothel named Le Petit Canard, but got absolutely no response at all. My North American bias put this silence down to French reserve. But when Lisa met Julie, she put her on the search. After digging and scraping, and after a long series of telephone calls, Julie finally got me an answer, though not the answer I had been hoping for.

Mousieur Pinette responded that the natural history collection in Amiens had not been accessible to the public since 1986, and had no curator. Between 1840 and 1940, a museum in Amiens housed a collection of 2,400 bird specimens, but that collection was largely destroyed when the town was bombed. At least one precious specimen had been spared the devastation—their stuffed Great Auk. There were about 600 bird specimens in the collection today, probably mainly from donations since World War II. As far as anyone knew, it was Great Auk, one; Labrador Duck, nil. My greatest hope was that I was going to discover the missing duck on a corner shelf that had been overlooked.

And then a truly wonderful experience began. To envision Pinette, imagine the actor John Malkovich. Now take away all of the implied threat of looming horrible violence and replace it with an incredibly hearty and disarming smile. Joining us was young Stéphane Herbet, who had been working for two years to organize the current incarnation of the natural history collection. We were welcomed as royalty. For Julie and me, this was a new experience in the comings and goings of a world-class art and culture museum. To Pinette and Herbet, it was insight into the world of the history of biology. We chatted loudly and proudly about the museum and my duck.

In the end, the story is distilled with a lot of unsatisfying residue. Today, the Musée de Picardie is concerned with art and archaeology. The history of the museum is poorly documented and the natural history component is just about the most clouded portion. For the twenty years leading up to 1986, the natural history collection had been on display in the museum on rue de la République, but had been removed to storage to make room for more displays of fine art. Until recently the Amiens zoo had owned the natural history material, which was then brought into the grander collection. Herbet was working to organize the natural history collection and create an inventory. Many items in the collection were not in the best condition, and consideration had been given to scrapping the whole lot. What was going to happen to the collection next was not clear. It had to be put to some use, but it was not Amiens' most urgent priority. While that decision was being made, preservation was the highest concern. We spoke about how the Great Auk might have survived the 1940 bombing when other specimens did not. We considered the possibility that it had been removed to safety at the start of the war, or that it had been retrieved from the flames by someone who recognized its value. It was all idle speculation, of course. Poor Julie was translating as fast as she could, trying desperately to keep up as the rest of us spoke, and I scribbled longhand notes.

As far as anyone in Amiens knew, there was no Labrador Duck in the collection, but we were very welcome to have a look if we wanted. I told Pinette that if I found a Labrador Duck, it would be a big story in the Amiens newspaper the following morning. When I told him of the value of a specimen, he asked if I could create one from bits and pieces of other specimens. I think he was joking.

The collection was housed at a site a few miles from the museum, and in an act of extreme generosity and bravery, Pinette loaned his car to Herbet to take us to that site. We pulled into the parking lot of an unmarked warehouse and entered through an unmarked door. Housed within we found a substantial collection of minerals and fossils, preserved and stuffed vertebrate animals, boxes full of pinned insects, and a mountain of pressed herbarium specimens. At first glance, it was clear why the rubbish heap had been considered as a destination, particularly for the herbarium specimens, which were in need of immediate emergency botanical attention. A more careful second look revealed a collection with many worthwhile treasures but one that could not survive forever in a warehouse.

The birds and mammals were housed on metal shelving, given some degree of protection by overhanging sheets of clear plastic. I examined all of the waterfowl in the collection very, very carefully, keeping in mind that Gurney might have made a mistake about the sex or age of the stuffed Labrador Duck he claimed to have seen. No Labrador Duck in sight. I then examined all of the other birds in the collection, in case something had been misshelved. Again, nothing, and this left me rather sad. It meant that the world didn't have fifty-four Labrador Ducks. We were down to just fifty-three.

Nonetheless, I took the opportunity to examine the Great Auk. It is a real beaut, particularly after the unhappy specimen in Strasbourg. It is the single best specimen in the collection, and if only one artifact were to be saved from the wartime bombing, how fortunate that it was the Great Auk.

So what can I conclude from my visit? Surely the simplest interpretation was that, in the past, Amiens had a museum with a Labrador Duck, as reported by Gurney in 1897, but the duck had been destroyed by bombing in 1940. End of the story? Herbet then showed me something that threw the story back into turmoil. It was a twenty-page document entitled "Catalogue de la Collection des Oiseaux du Musée d'Amiens," prepared by F. Choquart of the Northern French branch of the Linneaen Society, and published in 1897. It was a very thorough account of the 2,391 stuffed specimens of 518 bird species housed in the collection, from four Griffon Vultures, through eleven Marsh Harriers and twelve Ring Ouzels, to the single Great Auk. If you had to guess, was there any mention of the Labrador Duck described in the same year by Gurney? Not on your life.

The possibilities were now almost endless. Perhaps Gurney spied a stuffed Smew in a dimly lit museum room, thought it was a Labrador Duck, and mistakenly revealed this to the world. Perhaps Gurney really did spot a Labrador Duck and pointed it out to a member of the museum staff, who then sold it. Maybe Choquart missed the Labrador Duck in his inventory or chose not to include it for reasons known only to him. Take your pick, or make up a story of your own.

After lunch, a look at the cathedral, and a tour of the museum, it was time to catch the train back to Paris. We had found no new ducks, but had made a couple of new friends. I looked out of my hotel window at the Eiffel Tower in the distance before heading off into the night. On that last evening I found myself at a Paris café with sidewalk seating, drinking a beer whose name was thoroughly unpronounceable with my few scraps of French. It offered a hint of cloves and perhaps of nutmeg. I watched the passing of the early evening throng. Unlike London, in which the population generally seems confused, the people of Paris seem confident but star-crossed. Smiles often contained a hint of the ironic, and unleashed laughter seemed infrequent. Each person had the air of being largely in control, but seemed to know that fate has conspired to keep things from ever being absolutely perfect. In Greater Paris, some 470 persons had been born that day to replace a near equal number that had passed away. Somewhere in the great metropolis that night was a tormented soul who had tried unsuccessfully to end life under the wheels of the number 3 métro.

But then I cheered myself by thinking back to a particular mausoleum in the Cimetière du Père-Lachaise. It had been commissioned by the grieving Mr. Kennedy in 1856, who had just lost his twenty-six-year-old wife, Alice Emily Margarete, and his six-year-old daughter, Alice Maude. Instead of reflecting on the futility of lives whose threads were cut too soon, Kennedy decided to celebrate the positive impact of their brief dance with life. Engraved in the side of the mausoleum, in English, was a poem reflecting how I felt about my all too brief adventures in France:

There are days that might outmeasure years,
Days that obliterate the past,
And make the future,
Of the colour which they cast.

Chapter Six

The Invasion of Germany by Vandals

I was tired. I was grumpy. I had been working too hard and should have been resting. Almost as soon as my flight from Paris touched down in London, I found myself on a Lufthansa flight bound for deepest, darkest Germany, strapped in beside my English chum Errol Fuller, who was downing Bombay gin and tonics at an alarming rate.

I had two reasons for traveling to Dresden. The first concerned eggs. After the disappointing results of the genetic analysis of the eggs in Scotland and England, the only remaining possibility in terms of Labrador Duck eggs was to chase down six specimens that had been in Dresden before the war. The Staatliches Museum für Tierkunde was also blessed with a Labrador Duck hen that had survived the wartime devastation of Dresden. She was to be my twelfth. Paul Hahn didn't mention this duck when he put together his summary in the 1960s, and for a terribly good reason. At that time, the duck and other precious natural history artifacts were hidden in St. Petersburg, having been stolen by the Russian army at the end of World War II.

Given my successful massacre of the French language, I felt ready to try another tongue with which I had absolutely no facility: German would do nicely. Although willing to enter Germany without a proper minder, I wasn't willing to go without a traveling companion. I decided to have a go with someone equally incompetent with the language. Hence the presence of my chum Errol Fuller. Not only was he unable to speak any German at all, but I had a vague notion that he wasn't all that keen on Germans.

Errol, a self-described vandal, is never far from a little trouble. He likes to piss on rules, and would probably rather wear a brassiere than a seat belt. A look in his eyes speaks vaguely of danger. He seems quite taken with causing a bit of trouble here, there, and everywhere. If trouble requires a large gin and tonic, so be it. On our flight out of Heathrow, Errol's G&Ts looked so damnably good that I started gulping them down as fast as the flight attendants would bring them.

The Dresden airport is ludicrously spacious, roughly the size of Wimbledon—the city, not the tennis club. Two lovely and efficient ladies at the airport's information booth were waiting to provide us with any sort of assistance, and quickly booked us into a modestly priced hotel at the north end of the city, close to the museum, and arranged for a complimentary shuttle bus to pick us up. At the hotel's front desk, the incredibly helpful Rita provided us with a city map, along with a schematic and schedule for the city's tram lines and the requisite tickets for same. She then used the Internet to look up and call the natural history museum to find out exactly where we needed to go the next morning. Whatever Rita might be getting paid, she deserves a raise.

Eager to seek out new adventures, we hopped on the tram, which sped us into the heart of Dresden, where we immediately found a café, nearly empty, that served us some impressive beer almost before we asked for it. Hearing us speak English, our server whipped away the German menus and replaced them with an English translation. Always on the lookout for souvenirs, I asked our waitress if it was possible to purchase my beautiful gold-trimmed beer glass. She immediately came back with a clean one, wrapped it carefully in a newspaper, and told me the cost, which I missed entirely. Afraid to make a mistake, I produced a ten-euro note, hoping that it would be enough. She handed me eight euros in change.

Neither of us seemed to be troubled by a light rainfall, and so we let our dinner settle with a good long tramp across the river Elbe and through the town. Errol and I had been speaking all day about our professional experiences and journeys, but the night seemed better suited to talk of marriage and other relationships, of finances and children, of successes and failures, of goals and dreams. We found another nearly empty bar for yet another outstanding beer.

We gazed up at magnificent architecture that had survived wartime bombing, and a few ugly housing complexes presumably dating from the era before German reunification, when Dresden had found itself on the wrong side of the ideological fence. A couple of buildings, almost completely destroyed, had been left in place, perhaps as a remembrance of the devastation of war. Dresden seemed a perforated city, as though buildings had been destroyed, but after the rubble had been cleared away, nothing had filled in the gaps. Residents of Dresden claim to live in the most beautiful city in Germany, a bold claim in light of the nearly complete annihilation of the city resulting from the Allied firebombing of February 1945, followed by the construction of bleak factories and housing complexes of the German Democratic Republic era.

But I am on the side of Dresdeners, with a high opinion of their home. They live in a community most wonderful, as though it simply refused to let trifling matters like firebombing, flooding, and a communist regime get in the way of its magnificence. Its streets reflect a respect for the past, but also a firm belief in the tremendous potential of the future. Dresden is home to smart, well-stocked shops and trendy restaurants with keen servers. We had no trouble finding folks who spoke fluent English. The city is ripe for an invasion of tourists, awaiting just a few well-timed articles in the travel supplements of major newspapers in London and New York.

But I wasn't in Dresden just to see the sights. Six eggs and a stuffed hen were waiting for me at the Staatliches Museum für Tierkunde in the care of the curator of the ornithological collection, Siegfried Eck. Like so many of Europe's older natural history museums, this one has an interesting story. The collection dates back to the 1560s, and the museum became an independent institution in 1728. A fire in 1849 wiped away a good portion of its holdings, but it was rebuilt in the decades that followed. The museum was modernized and the collection expanded to take in great treasures, including more than 6 million specimens, with particular emphasis on vertebrate animals. Operating out of the central core, amid other components of Dresden's great cultural life, it must have been quite the operation.

But then administrators marginalized the collection. They closed the galleries to the public and moved the collection to a new building in the northern reaches of Dresden, close to the airport. Perhaps the natural history museum suffered as a result of Dresden's superabundance of thirty museums. Perhaps city planners thought a museum full of stuffed animals to be archaic. Like a chipped piece of pottery, still too good to throw away, but not good enough to put on display, it had been relegated to the garage.

Given that he had been publishing scientific articles since I was in junior high school, I expected Eck would be a frail old relic. Instead, a spry fellow, looking not much older than Errol and me, dashed down the stairs to admit us. He spoke less English than anyone else in Dresden, but was entirely hospitable, even if we were able to exchange only rudimentary pleasantries. He had prepared a workspace for me and took us to retrieve the eggs from a great locked cabinet behind two great locked doors. It was peculiar to see that after all the money that had gone into construction of a grand building to house the collection, the eggs were housed in a grocery store egg carton.

Settling into the workstation, I took out my tools for extracting material from inside the eggshells, donned my silly-looking helmet with magnifying lenses, and slipped on surgical gloves. If possible I wanted to avoid contaminating the duck DNA with my own. Our brisk walk to the museum and the overly warm building combined to make me start to sweat, presenting a problem. I didn't want to go to all this trouble only to ruin the egg material by dripping sweat on it. I repeatedly dabbed at my forehead with my shirtsleeve. Extracting material from the first four eggs went very well, and I was sure I had enough for Mike Sorenson to do DNA analysis, but the last two eggs were a bit trickier. The blowholes were absolutely teeny, and there wasn't much material inside to work with. Breathing through my ears so as not to blow away any fragments that I winkled out of the shell, I picked and poked with my surgical instruments. Part of me wanted to just grab a bit of the shell with forceps and yank, following my vandalism with an "oops!" but I resisted the urge. In the end, I had gunk inside six sealed tubes, and Eck had six intact egg shells.

Labrador Duck 12

Having been looted by the Russian army at the end of World War II, the well-traveled hen is safely back in Dresden.

Eck escorted us as I took the eggs back to their home, and traded them for the stuffed hen in her plastic tray. She was a pretty little thing, constructed as a taxidermic mount, but without a base. My best guess is that she was prepared to appear as though swimming, originally displayed with some elaborate system of support. The taxidermist had given her pale yellow glass eyes. Her tail feathers were a little beaten up, not unexpected considering her journey across the Atlantic, and then to Russia and back. Beside her in the cabinet was a test tube of body feathers that had fallen out in her trips to and from Russia. After examining and measuring her, I was able to add one more stray feather to the tube. That was it for the Dresden duck; there were now forty-one ahead of me.

• • •

THE NEXT MORNING we were off to a fortress in Königstein, where so many art treasures and valuable natural history artifacts had survived the 1945 bombing of Dresden. With typical efficiency, the number 7 tram took us right to Dresden's *Hauptbahnhof*, the main train station. As one of the last bits of city core to be revitalized, we found the station's front entrance roped off with blue and red barrier tape telling us: "*Wirten Was!*" Dresdeners clearly take their reconstruction very seriously, as numerous police officers and army personnel were stationed behind the tape, using their walkie-talkies. We found a side entrance, but it, too, was taped off. An ominous-looking green van with the word *Bundesgrenzschutz* on the side suggested that we were looking at more than just construction delays. We found a lady wearing the sort of uniform that a train conductor might wear in a children's book, and asked her what was up.

"They have found a bomb," she said, waving her arms enthusiastically. She seemed quite jovial about this bit of excitement, as one only can when being paid not to do proper work.

"Do you have any idea when the train station might reopen?"

"Five minutes . . . two hours . . . Who knows, who knows?" accompanied by lots more enthusiastic arm tossing.

"I trust that this sort of thing doesn't happen frequently?"

"Who knows, who knows?" I seemed to have found the limits of her English.

Standing around the train station waiting for the end of a bomb scare seemed an inefficient and possibly dangerous use of our day, and so we set off in search of alternative adventure. Errol sought out a soccer jersey featuring the local team, Dynamo Dresden, for his football-mad son, Frankie. We spent a couple of hours at the Zoologischer Garten and got to see naked mole rats. If you have never seen a picture of a *Nacktmull*, your imagination will not lead you far astray. They are the size of large mice, live in underground colonies, are nearly blind, pink, and almost completely devoid of hair. Imagine a scrotum with a head and legs.

From the zoo, a short tramp took us back to the train station, past rows of unlovely flats apparently scheduled for demolition. All signs of the earlier bomb scare had been neatly, surgically removed from the station. The lady who sold us our train ticket was very helpful—without being asked, she explained where and how to validate our ticket. Errol didn't want to validate our tickets, explaining that we could always feign ignorance with an incomprehensible accent from the southern United States, but I stuck them into the stamping machine anyway. Some vandals are a little more dedicated to rules than others.

No city shows its finest side to train travelers, and we watched a landscape of disused and unloved factories and Cold War–era apartment blocks, interspersed with people getting on with their lives. But the landscape changed profoundly as we reached the town of Pirna. From there, the train line followed the river Elbe for the remainder of the journey into Königstein. Passing through Obervogelgesang, Stadt Wehlen, and Kurort Rathen, we saw affluent homes with steep roofs, backed by tree-covered hills. The hills were then replaced by tree-covered cliffs, until the cliffs became too steep to support any trees at all. More level bits support broadleaf woodlands with a fern understory. Barges floated by on the Elbe, as did cruise ships. Sixty minutes after we started, we were at our destination.

IF I HAD tried to create a fairy-tale village from my imagination, I couldn't have done so nice a job as the 3,200 good people of Königstein. We wandered down narrow streets with narrow sidewalks, flanked by five-story buildings beautiful in their modesty, toward the town center. Parts of some buildings lay unoccupied, but these sat beside portions that housed fancy furniture stores and pharmacies. The Protestant church was undergoing loving reconstruction efforts. The town, sandwiched between the river on one side and grand cliffs on the other, at almost every point offers a view of the fortress looming above like a benevolent dictator.

To see the Fortress Königstein from any part of town, look up; look way, way up. You can't miss it. There are several ways to approach the fortress. The trip is quite easy if you are driving a car or belong to part of a tour group. We were neither. Alternatively, you can take a kitschy open-sided tour bus to the top. As vandals, we opted for a narrow trail that starts close to a bakery and runs up the hill to approach the fortress from the backside. If you choose this route, you will pass through dark forests that could easily be occupied by elves and fairies. However, if you have a heart condition or bad knees, please take the bus; the hill is steep and slippery and the cobbled surface tricky. On that particular Wednesday, we were the only vandals silly enough to attempt a rear attack on the fortress.

And then, just as we began to see the appeal of the kitschy bus, we came across the giant walls of the fortress Königstein, which translates as the "King's Stone." A breathtaking 790 feet above the Elbe, it not only provides a great view of the surrounding region, but makes you wonder how mad someone would have to be to try to invade the castle. In the early 1400s, a four-year siege was required to force a turnover; weapons of the time proved less effective than eventual starvation. In the following five hundred years, no one managed to beat these fortifications. At the end of World War II, it fell to Russian forces on May 9, 1945, without a single shot being fired. I suppose that massive sandstone cliffs and walls lose their protective value in the face of an aerial barrage.

Over the centuries, the fortifications at Königstein variously served as a military base with hundreds of soldiers, an oversized hunting lodge, a hospital, housing for an endless parade of political prisoners, a locale for court festivities, a camp for prisoners of war, and a residence for the fabulously rich. At one time it was home to the world's largest vat of wine, with a capacity of more than 62,000 gallons and room on top for thirty dancing couples.

The fortress was used to house Dresden's great works of art during the Seven Years' War in the mid-1700s, again during the Napoleonic Wars of the early 1800s, and then again during the Prussian-Austrian conflict later in the century. Dresden's museum curators, tiring of the back-and-forth movement of artworks, must have been just about ready to leave the whole lot in Königstein. Since the fortress worked so well to protect art during each of those conflicts, the treasures were moved back there once again during World War II. Starting in 1940, there were 450 crates of treasures moved from Dresden, down the river and up the hill to the fortress, and stored in artillery-proof vaults. Among the treasures housed in Königstein were the Dresden Labrador Duck, six Labrador Duck eggs, a Great Auk, and a Great Auk egg. On May 9, 1945, the Red Army moved in and took over the fortress, liberating prisoners of war. They also liberated the great treasures stored there, taking them back to Russia as spoils of war. Most of the artwork was returned to Germany in the 1950s, but some remains unaccounted for to this day. It wasn't until 1982 that Eck was permitted to travel to Leningrad, now called St. Petersburg, to retrieve the natural history artifacts.

An endless array of ramps and tunnels ended at locked gates. Errol and I started to formulate hypotheses about Dresden's missing works of art. In 1945, if we had been on the German side, responsible for the safety of great works of art, we wouldn't have been in a particular hurry to throw open overlooked chambers and shout, "Hey! Russian invaders! You forgot to take the paintings in here!" Could some art treasures still be hidden under the Fortress Königstein? We felt that if we had been provided with lock-picking tools, surveying equipment, and enough time, we might find some of the art treasures that were never accounted for after the war.

OF COURSE I have been setting you up for a punch line, and here it is. As soon as I returned from Germany, I sent the egg material to Michael Sorenson. He completed the genetic analysis and quickly got back to me with the results. Regrettably, I have to report that the six eggs in Eck's care in Dresden are also not Labrador Duck eggs. Instead, these eggs were produced by a Red-breasted Merganser. This is a perfectly nice fish-eating river duck, but nothing out of the ordinary. And so, to the best of anyone's knowledge, there are no Labrador Duck eggs anywhere in the world.

One of the peculiar things about scientific investigation is that one result isn't better than another. An outcome is an outcome. As a scientist, I wasn't really supposed to care about what Sorenson's analysis had demonstrated about the nine putative Labrador Duck eggs. And yet part of me felt that humankind had lost something. I had proven earlier that the world knows nothing about the nests of Labrador Ducks, and now I had shown that we know nothing about their eggs. The loss might be small and intangible, but the Labrador Duck enigma had become a little more profound.

At Dresden Airport, Errol and I cleared security and checked out the duty-free shops. I picked up a newspaper to see what it had to say about the disturbance at the train station the day before. The front-page headline read: "*Bomben-Fehlalarm bremst 45 Züge aus: Dresdner Hauptbahnhof wieder evakuiert / Auch diesmal keine Video-Bilder,*" but that information just left me a prat pretending to be able to read German. Another newspaper showed a picture of the roped-off train station beside a picture of a coffee maker. Perhaps the army had arrived to detonate a suspicious-looking package, only to find that they had blown up a perfectly innocent kitchen appliance.

Chapter Seven

Into the Mouth of the Tourism Dragon

As tourism superpowers, cities and countries come and go. At one time, anyone who was anyone must have gone to see the Hanging Gardens of Babylon and the Colossus of Rhodes. The flow of politically incorrect tourists to Spain to watch bullfights is drying up. Not so long ago, any young person with a backpack and a few dollars in their pocket just had to go to Australia. This was helped by the popularity of the first *Crocodile Dundee* film; the second *Crocodile Dundee* film ended the honeymoon with Australia just as quickly. With the popularity of the *Lord of the Rings* films, New Zealand became the destination of choice. However, when it comes to real staying power as a tourist destination, Ireland must rank at the top of the charts.

Travel guides offer long lists of reasons why Ireland is so popular. They claim that the island is tranquil and the pace of life relaxed. The scenery is varied and beautiful, the food delicious, and the people are easy to meet and courteous by act of parliament. Ireland is the single best place to go for no-fail fishing, challenging golf, unspoiled beaches, unparalleled sites of historical interest, and the world's finest crocodile wrestling. In a world jam-packed with everyday annoyance and irritation, Ireland must surely be one version of Heaven on Earth. Just two days after finishing my adventures in Dresden, I was drawn to Dublin by a Labrador Duck, lucky number 13. Lisa was drawn to Dublin by a Physiological Society conference at Trinity College, and I thought it a good chance to tag along as a spouse to take advantage of cheap university accommodation and continental breakfasts.

A pretty good chunk of visitors to Ireland will get their first impression of the country at the Dublin airport. This is a profound shame. At an international airport, the first encounter is always, of course, with crabby people at passport control, and those in Dublin are significantly crabbier than most, as though suffering from a massive collective hangover. Lisa and I got in a line with a big sign indicating that it was for non-EU travelers only. When we got to the front of the line, a person of superlative crabbiness told us, in a manner of speech reserved for dealing with those with serious attention disorders, that the line was reserved for EU travelers only. We were instructed to get in a much longer, much slower line.

Everything wonderful that Dresden airport is, Dublin airport isn't. The arrivals section was ridiculously dark, hellishly crowded, and generally uninviting. Conveyor belt 4 was blanketed in darkness except for a single naked 8,000-watt fluorescent tube that required us all to shield our eyes while watching for our luggage. The luggage-handling system would have been adequate for the bags of all four passengers disembarking from a Cessna 175 single-propeller airplane, but was laughably inadequate for the hundreds of passengers getting off our jet. Regrettably, conveyor belt 4 was also receiving the luggage from three other flights that had just arrived. Lisa stood well away from the crush, while I did my best to help a gaggle of elderly ladies who were having absolutely no luck swinging their luggage off the conveyor. I joined in their jokes about making do with the first piece of luggage that looked promising, until I realized that they weren't joking. My luggage is nothing posh, but I outdid one person using a pillowcase and a piece of rope, and another using a clear plastic duvet bag. One traveler's luggage consisted of a folding lawn chair held closed by packing tape. The tape broke, and the chair jammed the conveyor belt, holding back a tide of luggage belonging to eight hundred other passengers.

Dublin is home to about a million people, and is Ireland's capital city. According to guidebook author Catharina Day, its beggars, housing ghettos, and problems with theft and drugs, are offset by Dublin's "wealth of Georgian architecture, a lively, youthful atmosphere and a charm that is particular to the city itself. . . . Dublin," she claims, "has a worldwide reputation for culture, wit, friendliness and beauty," but then goes on to admit that "there is no doubt that Dublin can be a bit of a disappointment . . . Fast-food signs and partially demolished buildings mingle with expensive and tacky shops, and the housing estates can be depressing." The airport bus took us past a good chunk of the disappointing and depressing side, and we were looking forward to enjoying the lively and beautiful bit.

Luckily for us, Trinity College is an oasis of calm in a sea of turmoil. It was founded in 1592 but remained closed to Catholics, women, and other heathens for nearly four hundred years. Happily, the college is now open to all. Entering through an eighteenth-century facade, we left behind Dublin's hustle and bustle and entered a world of grass and cobblestones, separated from the real world outside by impenetrable tall buildings.

AT 10:00 THE next morning, Lisa was one of hundreds of physiologists at the opening sessions of her conference. At 10:00, I was one of nine people waiting outside a wonderful old stone building on Merrion Street for the opening of the Museum of Natural History, one branch of the National Museum of Ireland, also known as Ard-Mhúsaem na hÉireann. Lisa, the smart one in our marriage, had a presentation to make, entitled "Elevated [K^+] Enhances Cultured Adult Rat Cardiac Myofibroblast Contraction." I was going to play with a stuffed duck.

In the late 1950s, Paul Hahn's request to the National Museum of Ireland for information about stuffed specimens of extinct birds drew a response from Geraldine Roche of the museum's Natural History Division. She indicated that the museum had one Passenger Pigeon, two Eskimo Curlews, two Carolina Parakeets, some Great

Auk bones, and, most importantly, a Labrador Duck. She told Hahn that the duck had been brought from New York in August 1838 by Lt. Swainson, R.N., who was in command of the steamship *Royal William*. This ship had been in the service of the City of Dublin Steam Packet Company. The records do not say where Swainson got the duck, but they do say that he passed it on to Reverend Palmer Williams. When the good Reverend Williams passed away, the duck came into the possession of his niece, Mrs. H. M. J. Barrington, of Edenpark, Dundrum, Dublin. Before his retirement from the post of Senior Technical Assistant at the National Museum, Patrick O'Sullivan had written to tell me that the museum had purchased the duck from Mrs. Barrington on August 11, 1892, for the sum of £30, which was, no doubt, a lot of money at the time.

When I introduced myself at the front desk, a gentle little fellow buzzed Nigel Monaghan, Keeper of the museum's Natural History Division. As we walked through the public galleries, Monaghan told me that the museum has about 2 million specimens in total, eight to ten thousand of which are birds. At the time of my visit, all of the bird specimens were being added to a computer database so that researchers around the world would be able to access the information. The museum hadn't always been so committed to its birds. Damien Walsh of the Education and Outreach Department had told me that the museum had not had a curator of birds, or even a curator mildly interested in birds, for nearly one hundred years.

The museum opened in 1857, two years before the publication of Darwin's *Origin of Species*. Its opening was commemorated by a lecture by Dr. Davis Livingstone about his African adventures. Today, the museum is crowded and musty, but in an entirely appropriate sort of way. The grand old Victorian cabinets cannot have changed much since the museum opened. Admission is free, my favorite price for a museum. Exhibits are set out over four floors; the top two are encircling balconies that give the visitor unique perspectives on some of the larger displays below. The whole effect is bright, cheery, and grand without being too obvious about it. If you should wish to see one of Europe's very few Labrador Ducks on public display, you will have to travel to the Dublin Natural History Museum on Merrion Street, where you will find him in cabinet 230, on the third level of exhibitions, between a pair of King Ducks, above some eiders, and below a female Bufflehead.

Labrador Duck 13

On that particular day, the Labrador Duck wasn't on display but in a back workroom, waiting for me, perched on his two-tiered wooden base. It normally has a 20-inch-tall glass dome over the top, which probably explains why it is in such tremendous condition, despite having been on public display for as long as anyone can remember. For a long-dead duck, he looks very attentive. Someone had done a rather unfortunate paint and varnish job on the bill, making it pink with splashes of yellow and lime green. The bill is slightly darker on the left side, perhaps because that is his display side and has received more exposure to the light. His right leg is a bit bashed up, but otherwise he is in remarkably good shape.

By the time I finished my examination of the duck and waved goodbye to my host, the museum had filled with enthusiastic visitors of all ages—pleasantly occupied without feeling cramped. This tells me that this particular museum should not give in to the temptation to modernize it. Even though visitors that day were not able to see the Labrador Duck, they did have the opportunity to see an amazing array of taxidermic mounts, including a Carolina Parakeet, a Passenger Pigeon, an Aukland Island Merganser, a Slender Bush Wren, and a Huia, all extinct. There is lots of nonbird stuff, too, but I confess to something of a bias for things with wings.

Even though we don't know exactly where the Dublin duck came from, or who had it before it left New York, we have a complete record from 1838 onward, which is not bad for a Labrador Duck. We know about Lt. Swainson, R.N., who got it from New York; Reverend Palmer Williams, who got it from Swainson; and Mrs. H. M. J. Barrington, who got it from Williams and then turned it over to the museum in Dublin. Oh yes, and the steamship *Royal William*. With very little digging, I found that the *Royal William* had been 144 feet long, 26 feet broad, 16 feet deep, weighed 785 tons, and could generate as much power as 540 motivated horses. According to my sources, she made only three round trips across the Atlantic between

Liverpool and New York, in 1838 and 1839, but one of these trips had set some sort of record for speed of passage. Despite my best efforts in the Heraldic Museum and Genealogical Office, I learned nothing more about Lt. Swainson, Rev. Williams, or Mrs. Barrington.

While I was having fun looking at ducks, Lisa also was having fun, pushing back the world's ignorance about all things physiologic. While I was poking at things long dead, she was trying to keep folks from dying prematurely. In the spirit of mixing science with pleasure, the conference organizers had planned a mixer at the National Gallery for that evening. Opened to the public in 1864, the gallery boasts many hundreds of works of art, with representatives of all major European schools. I arrived a little before Lisa, and staked out a spot near paintings of naked women with large breasts, and took advantage of the young men and women with elegant white gloves who brought around trays of wine. I was just a tad sozzled by the time Lisa arrived.

I was a little more soused by the time we joined a tour of the facility given by the lovely Shannon, who took us to some of her favorite pieces. She was particularly keen to show us works by Rubens, Monet, and Picasso. I found something vastly entertaining in the combination of great art and glasses of free wine. The tour ended with one of the gallery's most prized pieces, the recently acquired *The Taking of Christ*. We were told that the painting had resided in the dining room of a Jesuit monastery in Dublin until 1990, when a painting restorer twigged that it had been painted by Caravaggio four hundred years before. Fueled by wine, I tried to bait Shannon. I asked if the painting was signed. No, it wasn't. So, I asked, if it had managed to sit unrecognized for all this time, what were the chances that it hadn't been painted by Caravaggio at all? No, no, Shannon assured me, it was a real Caravaggio. No doubt about it, she said. Absolutely sure.

Back at the main reception area, large, dangerous packs of physiologists had formed, and it was time to wander into the night to find food to help sop up a bellyful of wine. Lisa rounded up a group of eight physiologists who she felt might be willing to talk about things other than physiology, and we wandered west through the Temple Bar area in search of a suitable establishment. Temple Bar is a model of urban regeneration, having been converted from rundown dockland to a center for culture and entertainment. According to the book of matches I found in my shirt pocket the next morning, we settled on a restaurant that described itself as the ideal venue for large parties, graduation celebrations, and corporate functions. It proved an ideal venue for getting a drink in my hand without delay. I looked around the table through something of a haze. Besides Lisa, my dinner companions were Tony Blair, Judy Jetson, Papa Smurf, Alanis Morissette, Sir Bob Geldof, Queen Gertrude of Belgium, Thomas Hardy, and Ed Sullivan. I was seated between Judy and Queen Gertrude, and took no exception to this whatsoever. As we drank and ate and drank some more, the conversation drifted around topics like global politics, global warming, art with breasts (both paintings and sculptures), programmed cell death, the nature of Canadian cuisine, and the relative merits of beer in Ireland, England, and Wales. Lisa steered me back to Trinity College.

A RADIO ADVERTISEMENT from the Ireland tourism board told me to expect a "surprise every day." Having measured my duck, I had a full day to explore, and I really, really wanted to be surprised. Pleasantly if possible. I didn't think surprises would find me if I stuck to sites listed in tourist guides, so I decided to do a little exploring. I wanted to find some modest Irish treasure, something beautiful or romantic or tranquil that I could look back on in my old age and sigh. I reasoned that sensible cities are built on rivers, and that many of the most sensible cities are situated on rivers where they empty into the sea. And so, in an attempt to find what lay beyond the rather narrow limits of my Dublin tourist map, I headed north from Trinity College, aiming for the river Liffey and O'Connell Bridge. I had been told that O'Connell was the most important of the many bridges that cross the Liffey, and as good a spot as any to start my adventure.

I found hell, which certainly was surprising. An endless stream of heavy lorries raced westward, perforated occasionally by commuter vehicles. The din echoed off the stone-fronted buildings to create a cacophonous crescendo. Pedestrians were not at all welcome, and the very few other pedestrians who braved O'Connell Bridge looked as disoriented as I felt. Looking for inspiration, I stood beside a statue of Daniel O'Connell, and felt sad that I had no idea who he had been. Perhaps I should have continued north along O'Connell Street, which has been described as Dublin's answer to the Champs-Elysées, but I didn't. I might have marched up Moore Street and down Moore Lane, but I didn't do that either. Instead, I tossed an imaginary coin and started east along the river.

It is said that Dublin has leapt forward in the last fifteen years, after years of neglect, and that restored old buildings and redeveloped areas had replaced lots that previously sat derelict. When it comes to the region bordering the north and south shores of the Liffey, the city still has a long way to go. I found no trace of soul. I did, however, find a trace of Labrador Duck in the guise of an unlovely building with a plaque indicating that it had served as the original headquarters of the City of Dublin Steam Packet Company, whose ship *Royal William* had brought the duck to Ireland.

The region along the Liffey was rich in promises of redevelopment of the portions of the waterfront no longer used for shipping. Blocklong signs on plywood fences described luxury housing development, but at that point in its genesis it was still a load of broken concrete and rusting machinery, and I was reduced to dodging heavy truck traffic where sidewalks had been demolished but not replaced. Around a boat basin, I found a series of partially occupied warehouses whose businesses had nothing at all to do with water. Moving further and further east, I finally came to the district still used for real shipping. Large wire fences kept me from getting anywhere near any real ships.

Zigzagging back along the river's north side, I finally found something that touched me. It was a series of statues of emaciated figures in remembrance of those who suffered and died as a result of the Great Famine starting in 1845, and the lack of assistance that followed. In particular it was a tribute to those Irish families that had emigrated to Canada, and so helped forge the nation as it exists today.

THE NATIONAL MUSEUM of Ireland has an incredible archaeological collection, including material from the Stone Age onward. This is all housed in a magnificent building opened in 1890. The museum was hosting several special events and lectures through the summer. On the books was a special presentation on bats by Dr. Claire Cave, an example of nominative determinism almost as good as a bat biologist friend of mine who married a woman named Robin. The museum was also sponsoring a children's art competition on the topic of Ireland's watery places. And that very evening, I was due to give a lecture on Labrador Ducks.

I like an audience, and have no objection to helping with advertising. I once did a radio interview promoting a talk while suffering a blinding migraine, with a corn snake wrapped around my neck. I hadn't been asked to do any promotion for the talk in Dublin, except for providing them with a brief description. When Lisa and I arrived at the museum for my talk, we were surprised to see no posters, no bulletins . . . not much of anything. Indeed, the only advertising we found was in the museum's booklet of summer events. It was buried on page 38, in blue ink on a slightly lighter blue background. And so I wasn't altogether surprised to find that my audience consisted of just twenty people, including Lisa and two people who had helped to organize the talk.

Despite the relatively small group, I danced and pranced, told Labrador Duck stories, and threw in a plug for the National Museums every chance I got. After the talk and the question period, one particularly well dressed, bejeweled, and enthusiastic fellow came up to ask additional questions and make supportive comments. His name was John McKenna, and he came across as someone who felt that life was a really good bit of fun. When we were tossed out of the museum a few minutes after my talk so that the building could be locked up, McKenna and his partner, Trish, invited Lisa and me out for a drink at a watering hole just down the street. We politely declined, claiming that it had been a long day. McKenna asked again, promising that it would be "just the one." Again we declined. When McKenna asked again, we knew that there was no way to politely decline a third time. And so off we went to the Horseshoe Room at the Shelbourne Hotel.

The decor of the bar has been described as understated Gatsby. I suppose this means lots of mirrors, wood, and brass. A Dublin landmark for fifty years, we were told that author James Joyce had frequented the Horseshoe, although I suspect he didn't do a lot of his best imbibing there, having died several years before it opened. Lisa and I were dressed nicely, but I still felt out of place, surrounded by lawyers and parliamentarians in suits costing more than my first car. McKenna and Trish were able to put us immediately at our ease. McKenna had whiskey, Trish and Lisa had wine, and at McKenna's suggestion I supped a Guinness. It is said that Guinness doesn't travel well, and so this was my opportunity to drink it with the least amount of undesirable travel, having come from just down the road at St. James Gate.

Trish was willing to sit back and look lovingly while McKenna held court. We wandered all over favorite topics, including jewelry, art, travel, history, and archaeology. McKenna described himself as the "fecker" who was costing the Irish government millions of euros by preventing them from running highways through areas of great archaeological significance. We pulled out some stationery and a pen so that McKenna could illustrate some fine points of Irish language and geography. After just a few minutes, the paper looked as though it had been attacked by a toddler holding a pen for the first time, which McKenna put down to his dyslexia.

Just before the evening came to an end, McKenna signed over to us the copyright for an image that was going to make us rich. He scribbled it out in pen on another piece of paper. It was a stylized sketch of James Joyce, incorporating Joyce's eyeglasses into the word *Bloomsday,* a festival held on June 16 each year to celebrate Joyce's greatest novel *Ulysses*. I gather that if Dublin ever uses this logo, Lisa and I will be able to claim that we hold the rights to it, and sue the living daylights out of the city.

BEFORE FLYING OUT of Dublin, I had one more opportunity to find a bit of the city to adore. The opportunity came in the form of a pre-breakfast run. My earlier impressions about traffic noise had come at rush hour. Surely those impressions had been misplaced. Not so. Passing through the gates of Trinity College, I heard church bells chime 6:00, and yet the din of traffic was already at full roar. I hadn't given up on my thesis that the best of a city should be somewhere near its biggest river, and so I set off, running west along the Liffey. I crossed at O'Connell Bridge, and again at Ha'penny Bridge. Then

I ran across the Millennium Bridge, followed by the Grattan Bridge, the O'Donovan Rossa Bridge, and the Father Matthew Bridge. North on the Mellows Bridge, south on the Blackhall Palace Bridge, north on the Rory O'More Bridge, and south on the Frank Sherwin Bridge. At this point I ran out of bridges, and so ran back to Trinity College, having failed to fall in love with Dublin.

I confess that my lack of passion for Dublin is probably completely my own fault. We were in the city less than four days. I met only a handful of its residents. I didn't get to Phoenix Park or Dublin Zoo. I didn't take in Dublin Castle or the National Botanic Gardens or the Museum of Childhood. Instead, I tried to find beauty by walking and running along the river Liffey. I had walked beyond the limits of my tourist map. Tourist maps are designed to keep tourists away from grotty building sites, heavy truck traffic, and unsympathetic architecture.

So here is my compromise: if the Dublin tourist bureau were to invite me back for a longer visit, I would stick to the itinerary, and I would promise to enjoy myself. I would attend only preapproved festivals and take only guided tours. I would then write a glowing and complimentary article about the city at the end of my stay. I would be perfectly pleased to fly economy but would not say no to a room at the Shelbourne Hotel.

Having made a careful examination of more than a dozen Labrador Ducks, I thought it was a good time to see where some of them had shuffled off.

Chapter Eight

A Traveler's Guide to the Smells of East Coast Canada

I am not a hunter. Indeed, I don't know any vegetarians who are. In the past, quite a few people hunted Labrador Ducks, even though they reportedly tasted awful. The story goes that when a Labrador Duck carcass was brought to market in New York, it would hang there until it rotted off the hook, because everyone knew how truly dreadful they tasted.

Labrador Ducks presumably spent the summer breeding somewhere in northern Canada—Labrador might be a good guess. Although there is not a single record of anyone having ever killed a Labrador Duck on its breeding ground, it is safe to assume that it happened, but no one bragged about it. Labrador Ducks spent their winters in the vicinity of New York City. People killed lots and lots of these ducks on the wintering grounds around Long Island. It follows that those ducks must have flown between the breeding and wintering grounds, following the coast in both directions, and there are a few records of Labrador Ducks being shot while they were taking a short break on migration. Stopping for a snack and a rest, my poor little ducks were blasted away at by the citizens of New Brunswick and Nova Scotia. Having spent the summer tracking down stuffed ducks in Europe, I had just enough time for a visit to eastern Canada before getting back to my other duties.

I knew of three scenes of Labrador Duck carnage. I had seen the adult drake in Liverpool that had been blasted while it rested in Halifax Harbour. The soul of the drake in Ottawa had been sent to heaven while visiting Pictou. I was soon to see a male in Chicago that had been unlucky enough to think that Grand Manan Island might be a nice place to have a nap. It was also time for me to see a live Labrador Duck.

Waiting for me at Halifax International Airport with a smile and a warm hug was my dear friend Sarah Shima. In imagining Sarah, think of a slightly older and slightly less Goth Christina Ricci. Think of a lustrous china doll with a permanent, disarming smile, but a look in her eye suggesting evil thoughts. To share a few duck adventures with me, Sarah had driven all the way from her home in Knowlesville, New Brunswick, to Halifax, Nova Scotia.

We had a bit of time to kill before our first appointment, and zigzagging among the lakes of eastern Nova Scotia, we stumbled across a golf and country club. My dress pants and shirt and gray hair must have carried a bit of weight, as we were able to waltz into the bar and were served drinks on the deck without being asked to justify ourselves. And so, with the smells of golf course fertilizer and herbicide wafting over us, I explained to Sarah why I had asked her to accompany me on duck adventures around Atlantic Canada.

Now it was one thing to arrive in Halifax, stroll down to the harbor, and say, "Gee, so that's where they shot a lot of innocent Labrador Ducks." It is quite another matter to make a big deal out of it. It seemed to me that I needed to speak to the final authority about Halifax Harbour—the Harbour Master himself.

Captain Randall Sherman had a neat gray beard and was wearing a tie with sailboats, but his shipshape office was surprisingly uncluttered by naval paraphernalia. He is a compact and engaging man, and we were welcomed warmly. Captain Sherman explained that he had been a master mariner for twenty years before moving to the post of Harbour Master. He is now responsible for safety and navigation and for things that might blow up in the middle of the night. The port is particularly sensitive about things blowing up after the ship *Mont-Blanc,* laden with an incredible cocktail of explosives, blew up on the morning of December 6, 1917, killing more than 1,900 people, crippling many more, and leveling great chunks of the city of Halifax. It was the largest human-created explosion before the nuclear age. Trying to keep everything in running order and avoid a repeat of the *Mont-Blanc* disaster, Captain Sherman and his team at the Port Authority have to keep track of some 2,400 cargo ships using the harbor each year, as well as their 15 million tons of cargo. On top of this are 3,000 yachts out of 5 clubs, 3 harbor ferries, visits by 100 to 140 cruise ships annually, and the comings and goings of the American navy. Halifax has a beautiful, wide blue harbor. It is 65 to 100 feet deep and doesn't need dredging. It doesn't freeze over in winter, has no currents to speak of, and has comparatively small tides. All in all, this is a great harbor, as long as you are not a migrating duck.

I told Captain Sherman about the Labrador Ducks that had used his harbor until some 150 years ago. He explained that even at that time, my ducks would have found Halifax a busy port, although all of the vessels would have been under sail. In the era of Labrador Ducks, ships would pump greasy water out of their bilges, and toss all of their other waste into the harbor. Not so today. Spills are rare and dealt with quickly, and cruise ships have their own waste treatment plants.

This isn't to say that the waters of Halifax Harbour are crystal clear. There are 350,000 people in the Greater Halifax area, and 100,000 people on Halifax Peninsula proper. These people create an awful lot of waste, which is dumped untreated through 43 sewage outfalls, straight into the harbor. Some of these outfalls are in prime tourist areas. Millions of dollars have been spent studying the problem, and residents pay a surcharge on their water bills for improvements, but not a lot has been done to fix the situation. Despite this, an amazing assortment of wildlife uses the harbor, including minke and fin whales; mackerel; and blue, mako, dusky, hammerhead, and tiger sharks.

For a man responsible for 55,000 ship movements a year, Captain Sherman proved very generous with his time. Before leaving, I cautiously asked him about his relationship with *Theodore Tugboat.*

For many years, the Canadian Broadcasting Corporation aired a children's television program by that name. For fifteen minutes each day, *Theodore* and his fellow tugboats pushed other ships around the "Big Harbour" while learning valuable life lessons. Each episode was narrated by the only real person on the show, the Harbour Master, played by singer-turned-actor Denny Doherty. Sherman laughed at my question and said that the show had made him, or at least his job title, famous around the world. He had even appeared with Denny Doherty on a *Theodore Tugboat* Christmas special. Today, a full-scale recreation of *Theodore* sails up and down the harbour, providing tours for children of all ages.

As Sarah and I stepped from the Port Authority offices, *Theodore* chugged by. We waved at the people on board and they waved back. I contemplated a time when the harbor had been a regular stopover point for Labrador Ducks. It was a magical moment that was only slightly dampened by a smell that seemed very out of place. It took a moment to realize that it was the smell of raw human waste. We left Halifax for the two-hour shot north to Pictou and my date with the ghost of Reverend McCulloch.

YOU WILL REMEMBER McCulloch as the Presbyterian minister who is spending the better part of eternity in hell for shooting Labrador Ducks in Pictou, including the drake whose stuffed remains resided for a long spell in Halifax before coming to rest in Ottawa. I wanted to see where McCulloch had spent his predamnation days, and where he shot Labrador Ducks. And since I was going to be there anyway, I had agreed to give a Labrador Duck talk the following evening to the Pictou County Genealogical and Historical Society, with particular emphasis on McCulloch.

Sarah had booked us into a cheery and welcoming little bed-and-breakfast operating out of a house built in 1840. The owners gave me the impression that they really wanted Sarah and me to be a married couple, given that we were sharing a room to save money, and so I didn't hesitate to use the phrase *my wife* whenever I could slip it into a conversation. This probably left them wondering what a sweet young thing like Sarah would see in an ugly old geezer like me.

With time before supper, "my wife" and I ambled through the

town. We found no shortage of grand old homes with magnificent views of the harbor. Even the more modest homes had a sense of pride and maintenance. We discovered signs explaining that in Pictou the minimum fine for parking in the spot at the post office set aside for disabled drivers is $136.25, and that the fine for loitering in front of the bank after being asked to shove off is $155. We found that the official mascot of Pictou is Causeway Cory, a cormorant dressed in raincoat, rubber boots, and a sou'wester hat. At one time, Pictou had been supported by a big shipbuilding operation. Today the major sources of revenue are tourism, a facility building parts for oil tankers, and a large pulp and paper mill.

Immediately across from our B&B, the Historic Sites and Monuments Board of Canada had erected a monument commemorating Pictou Academy, which was linked to my current duck quest. In both French and English, a plaque told me that, on the site, in 1818, Reverend Thomas McCulloch had established the first home of the academy. The institution was modeled on the universities McCulloch knew in Scotland, and emphasized "logical argument, scientific practice and equality of educational opportunity." McCulloch was clearly an important and influential man. Even so, I would have thought that he could have come up with more practical subjects than logical argument to teach to the men and women of a small, early-nineteenth-century Canadian fishing village. Swimming for one's life after falling into icy water from a capsized fishing boat, for instance. It took me 792 steps from the site of the old Pictou Academy to reach the house erected for McCulloch.

That night, I was delighted to find that Sarah snored like a band saw. As an insomniac, I am always afraid that my tossing and turning will keep a roommate awake. No problem in Sarah's case; one minute after her head hit the pillow, from across the room, the serenade began. I inserted foam rubber earplugs, and tossed and turned to my heart's content.

The next morning, I rose early for a walk through Pictou while Sarah had a little lie-in. I headed uphill, expecting to find less grandiose homes further from the water, but found quite the opposite. The homes were palatial, with big porches and greats chunks of well-maintained grassy property around them. I wondered how much this sort of palace with a great view and one and a half acre of land would cost in my hometown, before realizing that no such combination existed in my hometown.

After breakfast, Sarah and I set off for the Hector Exhibition Centre, where we were greeted by St. Clair Prest, who had coordinated my talk. Prest is from Moose River Mines, which has the distinction of being the site of the first remote radio broadcast in the region following a mine disaster. I suspect that Prest had spent a considerable portion of his childhood explaining his given name. The original St. Clair was, in fact, a fellow who had died 1,400 years earlier, and had been designated one of seven patron saints of tailors.

We gabbed with Prest about the Reverend McCulloch and the comings and goings of Pictou. He told us that the rink of the New Caladonian Curling Club (established in the 1850s) had three sheets of ice, but, having been built on reclaimed land, it had settled oddly, making the flight of rocks a bit tricky. He also explained that Pictou is gripped by five months of substantial snowfall each year, taking the shine off the otherwise idyllic community.

McCulloch House, maintained as an historic site, was closed for renovation and restoration when we visited. Ever willing to please, Prest opened it and showed us around. He indicated that the second floor had been added since McCulloch's time. In its earlier form, it had just one floor with four great rooms, with a sleeping loft above. There were about a dozen stuffed birds on display from McCulloch's collection. There was also a beautiful painting of a "Labrador Falcon, *Falco Labradoria*" on display. It measures 58 by 86 cm and has a dedication in the lower left corner, which reads: "Presented to Thomas McCulloch Esq. by J. J. Audubon." It was truly beautiful with exquisite detail that I could appreciate fully only with a magnifying lens. I fear for this work, because the roof of McCulloch House has leaked at least twice in the past. The picture isn't in pristine condition, with little creases and some small staining from damp, but it will take only one really good leak to wreck it beyond repair. I suspect it is of tremendous historical value.

That evening, Sarah and I arrived at the Hector Exhibition Centre half an hour early for my talk about Labrador Ducks. I became a little bit edgy when we found only one car in the parking lot. It was

Prest's. I started to worry that I might be talking to a rather teeny audience. At least the "Open/Closed" sign was turned the right way around. Prest had found a slide projector in a back cupboard and hoped that it worked. Luckily it projected an image, even if it rattled like an old Triumph TR3.

I needn't have worried about the turnout in Pictou; the evening drew more people than my talk the month before in Dublin. It was quite a social occasion, and ladies brought in coffee and tea and trays of sandwiches with their crusts cut off. The audience was enthusiastic about my talk, so I was wildly enthusiastic in return. I used funny voices and told funny stories and jumped around like a lunatic. An impressive string of questions and comments followed. Leaving the Hector Centre, Sarah and I retired to a bar with a deck for a drink to unwind. I had two large beers, and Sarah had half of a small one, which, she said, made her lips numb.

We had been given every indication that the unrelenting winter weather keeps Pictou from being completely idyllic. When the snow arrives, it shoulders its way in like a drunken sailor and is reluctant to leave without a fistfight. Sitting on the deck of the bar, drinking my beer, I discovered a second small flaw in paradise. Our world is addicted to paper, and when you live in a country that makes a sizable portion of its revenue from cutting down trees, you can expect that some folks are going to make their living by converting wood into paper. Hence the small flaw. Pulp-and-paper mills stink. Pictou has a mill. It isn't a stuff-a-skunk-up-each-nostril kind of smell, but it isn't an orchid-corsage-on-the-dress-of-the-prom-queen odor either. Sort of acrid, sort of sweet, sort of like syrup and pork fat, it made me wonder how people working at the mill manage to deal with it.

THE NATURALIST Charles Darwin will always be best known for his ideas on natural selection as the driving force behind evolutionary change. He published these ideas in a book whose rather ponderous twenty-one-word title is usually abbreviated to *On the Origin of Species*. He is considerably less well known for having written on a number of other topics, such as corals, carnivorous plants, and earthworms. The work that concerns us here is a two-volume doorstop entitled *The Variation of Animals and Plants Under Domestication*.

I would be lying if I said that I have read the whole thing, but it would be only a fib to say that I have read chapter eight in volume one, which considers geese, peacocks, turkeys, guinea-fowl, canaries, goldfishes, bees, silk moths, and domestic ducks. In this chapter, Darwin explained that, in the nineteenth century, "Labrador Duck" was a name given to a breed of domestic duck, which was also known as the Canadian Duck, the Buenos Ayres Duck, and the East Indian Duck. Darwin kept Labrador Ducks at his home in Kent. It seemed pretty obvious that if I were ever going to see a live Labrador Duck, it would have to be the domestic version, most commonly known today as the Black East Indies.

It had not been easy finding a breeder of Black East Indies. They may have been popular at one time, and goodness knows there are a lot of domestic-duck enthusiasts in the world, but not a lot of them breed Black East Indies. In the end I found a breeder, Margot Morris, hiding away in the vanishingly small community of Riverside-Albert in New Brunswick. Morris came out to greet us as we pulled up her drive. She must be a septuagenarian, but she had the sort of sparkle in her eyes that I normally associate with glass lenses inserted as part of a cataract operation. She was demonstrative and laughed easily and was genuinely happy to see us. We met her donkey and her horse, who both looked thoroughly pleased with themselves. Welcoming us into her home, Morris introduced us to her blue heeler, Shadow, who was very keen to thrust a wet chew toy into the lap of my white trousers. Morris' budgie, Maxie, remained politely in his cage, keeping all of his toys to himself.

This menagerie was all well and good, but there wasn't a duck to be seen. I had been led to believe that something like ninety ducks, geese, and swans had been sharing Morris' life. It seemed that Morris was ready to call it a day after fifteen cold, wet winters in New Brunswick, the most recent of which had been an absolute nightmare, with one great dump of snow after another, temperatures that turned her house into an icebox, and a winter virus that had been unwilling to turn tail. She was going to move to Ontario to be closer to her family. She had sold off her waterfowl in preparation for the move. Hence, there were no Labrador Ducks/Black East Indies for me to see.

We did have a lovely hour-long chat about waterfowl in general, and Black East Indies ducks in particular. Morris explained that as newly hatched chicks, Black East Indies are black all over, except for a few white feathers on the breast. They develop a green sheen as they age. She said that "the little devils run fast." Morris claimed that she was not particularly fond of ducks, and then chuckled when she realized how strange that sounded, given her hobby. She said that ducks, unlike geese and swans, are flighty and nervous. "You can argue with a goose," she said. "You might not win, but at least you can argue with them." The same is not true for a duck. She explained that duck flesh tasted bad because of their diet, which included just about anything they could catch, and gave the example of Jim Blewett, a singer-songwriter, who had been picking one hundred slugs a day out of his garden. After purchasing a couple of ducks from Morris, his slug problem disappeared. As a result, he was working on a folk song about ducks and slugs. Something along the lines of "You can have slugs, or you can have ducks, but you can't have both."

In my email messages many months before, I had asked Morris to set aside a particularly valuable prize. Domestic ducks were derived from Mallards through selective breeding. That is why Darwin was interested in them. Thinking that the DNA extracted from a Black East Indies duck might give me some insight into the eggs analyzed by Sorenson, I had asked Morris to set aside some eggshell fragments when her ducks bred in the spring. Morris had done much better than that. Before selling off her menagerie, she had set aside three intact duck eggs. They had been sitting in her kitchen beside the stove in a Tupperware container for two months since she collected them. As we spoke, Morris worked her way through a pack of Black Cat cigarettes, while the eggs stared at me from the container on the kitchen table. They bobbed slightly in half an inch of evil green ooze. Despite the best intentions of the engineers at Tupperware International to keep freshness locked in, the container was emitting the unmistakable odor of horribly rotten eggs. Perhaps after years of chain-smoking, Morris had completely lost her sense of smell. I felt really badly for poor Shadow.

At the end of an hour, we thanked Morris, wished her the best of luck in selling her home and with her move to Ontario, and drove off with our little Tupperware prize. It was all Sarah and I could do to keep our breakfast in place. Even though it was raining, we drove with the car windows wide open. We had to do something quickly. In the most remote corner of a parking lot, I pried open the container's lid and started to heave. Imagine hiking through a sulfurous swamp for a week. At the end of your trek, take off your hiking socks, and fill them with Parmesan cheese and vinegar. Then have a baby puke on them. That is the smell of two-month-old Black East Indies duck eggs. I decanted the ooze, swirled the eggs in some fresh water, and decanted again. I resealed the container, wrapped it in a plastic bag, and put it back in the car's trunk to simmer.

We drove on to Saint John, the second-largest city in New Brunswick, and all too frequently confused with St. John's, the capital of Newfoundland and Labrador. Our hotel gave us a dripping faucet. Rather than let all of that fresh water go to waste, I turned on the bathroom fan, opened the egg container, and placed it under the drip, allowing the water to carry some of the stench down the drain. We abandoned our room and found a very nice Mexican restaurant with a flamenco guitarist.

Too afraid of the smell to return to the hotel room, we allowed ourselves to be drawn to the waterfront by the sound of music coming from that evening's incarnation of the Saint John Festival Summer Stage. In a waterfront bar patio, with the delicate scent of the evening sea washing over us, I ordered a beer and Sarah ordered a tonic water. The beer's plastic cup, the umbrella above us, the music stage, and two giant inflated beer cans were all courtesy of Alpine Bière Lager Beer, maritime brewed (Brassée dans les Maritimes) by Moosehead Breweries.

On stage, a group of four young people played hits by the Doors, Black Sabbath, Uriah Heep, Led Zeppelin, and Pat Benatar. They all seemed to be too young to drink the product of the sponsor. The bass player might have received his guitar as a present at his recent bar mitzvah. The lead singer's breasts seemed impossibly close together, and my second beer had me trying to figure out the hydraulics of her situation. The gathered crowd was sparse, made up mostly of friends of the band, also too young to drink legally, drinking surreptitiously at a frantic pace.

• • •

The next morning found us traveling south along the Trans-Canada Highway, along the coast and past dwarf coniferous forests toward the community of Black's Harbour. The ferry just beyond would take us to Grand Manan Island, another Canadian site of Labrador Duck slaughter. Or so it would seem. The Field Museum in Chicago was home to two Labrador Ducks that I would soon be visiting. The good folks at the museum are not at all certain where the ducks were resting when blown to kingdom come, but one of their best guesses is Grand Manan Island.

A little more certain was the killing of a hen Labrador Duck off the coast of Grand Manan by a gentleman named Simon F. Cheney. Indeed, having been shot on its northward migration in April of 1871, it was one of the last Labrador Ducks ever collected. According to Cheney, just before it shuffled off its mortal coil, the hen had been eating small mussels and had been diving with Oldsquaw ducks. Without preparing the hen taxidermically, Cheney passed the skin along to Harold Herrick, who then passed it along to George A. Boardman, who then passed it along to John Wallace of New York so that it could be prepared properly before finally passing it on to Professor S. F. Baird of the Smithsonian in Washington. But just as a rumor can get distorted as it passes from one person to another, a duck passed from one hand to another can go astray, as this one did. Not recognizing the value of the hen, Wallace let the skin get away from him, and it has been missing ever since.

Regrettably, we got to the Grand Manan ferry two minutes late for the 9:30 sailing. Sarah was philosophical about it and politely suggested that I might want to have a go with the eggs, which were, quite frankly, causing her car to be unridable. The tide was out and I took the eggs and Sarah's pocketknife down the gravel embankment to a spot where a rivulet of fresh water ran down to the ocean. I steeled my nerves and, following in the long tradition of egg collectors, but fearing an explosion, I poked a very small hole through the shell at the blunt end.

I suspect that more Christians would be frightened into good behavior if ministers described hell in terms of the smells that issued from that egg. Saving oneself for marriage would seem like the only possible course of action if the alternative was eternity exposed to that smell. But to draw a parallel of that sort would be to tarnish the good name of hell. The moment that I pricked the eggshell, a geyser of frothy green liquid shot from the hole and arced before hitting the ground more than a yard away. Seconds passed, and the geyser didn't stop; I started to fear for the local environment. Thirty minutes later, by poking and prodding and rinsing the egg under the rivulet, I had managed to get all of the contents out and leave the shell intact. After another thirty minutes of this disgusting display I had the second shell cleaned out. I didn't have the heart to repeat the performance and, with the ferry pulling back into its berth, I tossed the third egg on a rock. It exploded.

Grand Manan Island has all the wonder of a tropical island paradise, without all the worry about coconuts dropping on your head. It is about 17 miles long and 7 miles wide, slung low in the middle but rising to a peak with a lighthouse at either end. It features lovely long beaches and quaint fishing villages. After dropping off our gear at a B&B, we drove to the island's south end. Just before arriving at the lighthouse, we drove into a thick fog. In an attempt to repel unwary sailors, the foghorn sounded for five seconds out of every sixty, but some trick of the fog meant that the horn echoed for a further fifteen fading seconds. Walking along the bluff, we got intermittent glimpses of a crashing surf far below and imagined what the view would be like on a clear day.

Driving north, we stopped at a beach. I wanted photographs of a spot where Cheney's Labrador Duck might have been shot. It was all idle speculation, of course; one spot was as likely as another. On that day, the surf was filled with sea ducks, foraging over the sandy bottom as Labrador Ducks would have done. Edible mussel shells that had washed up on the beach told me what my ducks had been eating.

All along the East Coast, fisheries are in decline. However, fishing seems to be alive and well in the waters around Grand Manan. We saw fishing weirs and salmon farm enclosures and an assortment of fishing boats. Everywhere the island had a fair perfume, a mixture of kelp, sea air, and fish processing. Not at all unpleasant. Just kind of fishy.

• • •

We were up early the next morning, hoping to fit a lot of living into the day. Our B&B hosts were apparently quite devout, as evidenced by the painting of a young sailor at sea, with a much larger-than-life Christ looking over his shoulder. Once again we seemed to be in the presence of folk who badly wanted Sarah and me to be married. Not wanting to disappoint, I didn't do anything to convince them otherwise. When our hostess commented on how good it was of me to bring all of the suitcases to the car, I said, "Well, the missus has me well trained."

In an attempt to see as much of Grand Manan as possible, we aimed for the northern tip of the island. There we came across the Swallowtail Lighthouse, hoping to find it fog free. No such luck, the foghorn sounded for three out of every twenty seconds. Even so, the walk was very pleasant, with steep cliffs and periodic views of a crashing surf and large seals or small sea lions. Back in the center of Grand Manan, we meandered a beach in a drizzle, hoping to find some marvelous treasure. I found a shard of blue crockery and managed to convince myself that it was from a shipwreck of 1620.

By this time the island's two museums were open, and I was keen to see both. The first was the Whale and Seabird Research Institute Museum. For such a grand name, I expected a little more. There were a few stuffed birds, a skeleton of a minke whale, and an assortment of marine mammal vertebrae. In glass jars, we saw some preserved invertebrates, the eyeballs and kidneys of a seal, and the dung of a right whale. It might easily have been the vomit of a right whale, and I had to wonder who had collected it, why, and how.

The second venue was the appropriately named Grand Manan Museum. Looking quite worn from the outside, it turned out to be rather nice on the inside. It had displays of local geology, mineralogy, and island history, including fishing, churches, and shipwrecks, all closely intertwined. One of the odder items on display was the remains of a naked-lady masthead from a shipwreck. Whoever found the masthead was so incensed by the lack of modesty that he cut it off at the head and burned the torso. There was a very large display of stuffed birds from the collection of Alan L. Moss, 1881–1953, the "birdman of Grand Manan." An annotated catalogue of the Moss collection included no stuffed Labrador Ducks, which wasn't surprising as the duck went extinct several years before Moss was born.

In the museum's library I found Anneke Gichuru and said that I was interested in information about notorious duck killer Simon Cheney. She explained that the Cheneys were a renowned seafaring family that still lived on the island, including Captain Craig Cheney. Gichuru suggested that the best person to contact would be island resident Brian Dalzell, who knew as much about the birds of Grand Manan as anyone. When I reached him on the library's telephone, he asked, "Are you a birdwatcher?" "Well, sort of," I replied. "I'm an ornithologist."

I told Dalzell about the Labrador Duck quest and my search for Simon Cheney. He said that Cheney was something of a mysterious figure, well known to the naturalist community of the day, but largely unknown today. Dalzell was unaware of any photographs or written records. He added a slight twist on the specimen shot in 1871, claiming that Cheney's stuffed-bird collection had been purchased by the owner of a general store on Grand Manan called Newton, who recognized the value of the Labrador Duck and sold it. In any case, the specimen is missing.

A crush of cars was trying to leave the island, and we were lucky that Sarah had booked a spot in advance. Our second sailing through the Bay of Fundy was as calm and foggy as our first. We saw many dozens of porpoises and hundreds of shearwaters. When we got back to Black's Harbour, we stopped near a convenience store to use the phone booth. This would have been a much more useful exercise if there had been a telephone in it. We went into the store where the lady behind the counter explained that youths had ripped the telephone out of the box so many times that the New Brunswick telephone company now refused to fix it. I made what I thought was a very reasonable suggestion. If the box didn't have a telephone, wouldn't it be more sensible to take away the box so that people wouldn't stop mistakenly? No, I was told, because then people wouldn't come into the store looking for a pay phone there.

"Does the store have a pay phone?"

"No, it doesn't."

"Doesn't that irritate people?"

"Yes, sometimes, but they usually buy something anyway."

And so Sarah and I were on our way across New Brunswick to her home in Knowlesville. Like me, Sarah prefers back roads, and there weren't any highways that could have got us there any faster anyway. We passed beautiful lakes with a surprisingly small number of cottages. After getting through cottage country, we passed through sparsely populated countryside. Most towns had a church, a cemetery, and a few houses. Each town seemed to be populated by folks trying their best to scratch out a living, sometimes by means that you just knew couldn't possibly work. In the middle of nowhere, one person had set up an ice-cream stand, and a few miles down the track, another had set up a car customizing shop.

As we approached Knowlesville, I suspected that Sarah was taking a random series of turns. This wasn't overly disturbing, as we passed a lovely selection of covered bridges and many signs that directed the traveler toward even more of them. We passed through a community boasting the province's greatest number of churches per capita. Citizens of the community had apparently used a plebiscite to vote down the opening of a liquor store.

We did eventually get to Sarah's home, at the Falls Brook Centre in the New Brunswick wilderness. Falls Brook is a nonprofit organization devoted to sustainable development. The facility seemed to be populated mainly by young women working for room and board. Sarah's home was composed of a tiny kitchen-cum-dining-room-cum-living-room, a small bedroom, and a third room with a shower and sink. The outhouse was a short stroll down a grassy trail.

In the months leading up to my trip, Sarah and I had been teasing each other about going skinny-dipping. After all, we were hearty Canadians in the hinterlands of New Brunswick, and Sarah claimed to know a great swimming hole along a creek. And so after dinner, we took to the roads again. Back roads turned into very back roads, and I was convinced once again that Sarah was making up the route as she went along. Perhaps there was no swimming hole, and she was hoping that we would stumble across something appropriate.

But, good to her word, Sarah found the spot, and after a final unlikely turn and a short trip down an unlikely dirt track, we found ourselves next to a Quonset hut used in making maple syrup. It was a really dark night, and we used flashlights to follow a path to the promised creek. In the flashlight's beam we spotted crayfish, and when Sarah picked one up, it nipped her. Off with the clothes, I followed Sarah in the hopes that the crayfish would get her before getting me.

At this point, I have to defend myself by saying that I have participated in two New Year's Day polar bear swims in the north Pacific Ocean, and one in the northern reaches of the Atlantic. I know what cold water is. However, it was clear to me that the water in this creek had just melted from a previously undiscovered glacier about 50 yards upstream. By the time I was in up to my knees, my brain had lost all control over the muscles in my legs, which were shaking uncontrollably. With every step I took, I knew that Sarah was a few steps further, a few steps deeper, showing herself to be a few steps braver. She kept coaxing me on, but when the water got to within a couple of inches of my delicate bits, I had a decision to make. I could bail and risk being teased, or go further on and risk alienating my testicles, perhaps permanently. I bailed.

It was a really dark night, both moonless and cloudy. The fireflies didn't seem to mind, winking a cool blue light. I suppose if you are a male trying to get a little action by signaling to females with a flashlight on your backside, the darker the evening the better. Using great patience and a gentle hand, we caught a couple. At some point in the past, I had probably learned that fireflies aren't really flies, but beetles, but it came as a surprise to me on that night. Standing naked beside a woman who wasn't my wife, I was reminded of a predatory bug that mimics the flash of male fireflies. When the lady firefly arrives, the predator devours her.

Chapter Nine

Garage Sales and Other Lies

My grand quest was, of course, designed to allow me to have a little poke at every Labrador Duck in the world. A search for some of the ducks, or even most of the ducks, wouldn't be nearly so challenging as having a go at absolutely every single specimen, no exceptions, period.

The quest led me down some dark and strange alleyways. After hearing one of my talks, a woman named Sandra Kaufman told me that she had seen a stuffed Labrador Duck in the post office of the tiny community of Ilulissat, Greenland. She said that the store was in the northeast corner of the town's main intersection. The duck itself was on a high ledge on the east wall of the store toward the back. Before shelling out a good chunk of my year's income on a journey to somewhere a little beyond the back of beyond, I sent off a couple of letters. Less than a week later, I heard from Ulf Klüppel of the Ilulissat Tourist Services, who told me that the bird Sandra had seen was an eider, not a Labrador Duck. Ulf's message saved me a long and costly journey, followed by crushing disappointment.

Not all leads were quite so easy to follow. In 1992, Stanford University Press published a tidy little book about extinct and endangered birds, written by Paul Ehrlich, David Dobkin, and Darryl Wheye. The

Great Auk, Passenger Pigeon, Carolina Parakeet, and Labrador Duck each have their lives summarized in six brief paragraphs, one each about their food, their nesting habits, their range, and so on. The same treatment continues for not-quite-extinct-yet species, such as Bell's Least Vireo, the California Condor, and the Whooping Crane.

For the Labrador Duck, under the heading "Notes," the authors state: "Until recently, it had been assumed that there were 31 specimens in North American collections . . ." which is reasonably close to the truth, ". . . but recently another one turned up at a garage sale!" Having been to a few garage sales, I have rarely found anything worth opening my wallet for. It really annoyed me that someone else might have found something as valuable as a Labrador Duck, and probably picked it up for about a buck and a half. Keep in mind that no new stuffed ducks had been discovered since the late 1940s. If such a specimen existed, then I had to see it. No exceptions.

I went to my copy of *The Big Book of North American Ornithologists* and found the postal address of the book's senior author, Paul Ehrlich. I wrote asking him to fill me in on this "new" specimen. He ignored me completely.

So I wrote to the book's second author, David Dobkin, again asking after the garage sale duck. He joined his colleague in enthusiastically pretending that I didn't exist. I couldn't find an address for the third author, and working on the theory that if a fly buzzes you often enough, you have to slap it, wrote to Ehrlich again. He slapped me. He returned my letter with the following note scribbled at the bottom: "sorry—no longer have a clue try David Dobkin." I did, and was ignored as thoroughly as the first time.

Then, three months later, I got an email request from the editor of the academic journal *The Condor,* asking me to review a research manuscript. Academics are asked to review manuscripts all the time as a professional courtesy, and refusing to help is considered to be bad form. Hence I immediately responded that I would be pleased to help out and then sat staring at my computer screen. My brain said that I should be able to recognize something about the message from the journal . . . something vaguely familiar . . . My brain said, "I need a better hint." Perhaps it is something about the origin of the message . . . "Nope, try again." Maybe I should have another look at the name of the editor of the journal, the fellow who sent me the message. . . .

There it was! The editor was David Dobkin, the extinct bird book's second author who had been ignoring me so effectively. I figured that I had him now. How could I be so polite in agreeing to do work for the journal and have him ignore my simple request for help? I quickly dashed off a second email message, reminding him about my query about the garage sale duck, and immediately got a huge and enthusiastic response. He suggested that I might want to try the third author, Darryl Wheye, who was living in California. I got a telephone number from the Internet, made a couple of false starts, but finally caught her, out of breath having just finished a bicycle ride. She got right on the case.

Wheye found that the comment about a garage sale duck had been provided by James Tate Jr. of the U.S. Fish and Wildlife Service. I contacted Tate, and although he couldn't remember the story exactly, he thought that it was based on a conversation that he had had many years before with John P. Hubbard, Bob Tordoff, Robert Storer, or Norm Ford while they were all at the University of Michigan in Ann Arbor. As Tate remembered it, he and Hubbard had been at the university's Museum of Zoology discussing some recent interesting developments in ornithology. Hubbard told Tate about a fellow graduate student who had found and bought a stuffed Labrador Duck at a garage sale in Ann Arbor and turned it over to the museum. "Apparently my memory is faulty," wrote Tate.

Tate was in touch with each of his former colleagues, and claiming that his big mouth had gotten him into trouble again, asked if they could remember the conversation from so long ago. The responses came back one at a time. No. No. No. And then, fully two years after I started searching for the garage sale duck, the answer came from Robert Storer. "I think I have the answer to your questions," wrote Storer, "although it is surprising how rumors have altered the facts." A taxidermic mount of an adult male Labrador Duck, it was in a group of specimens that had been lying neglected in the University of Vermont's small museum. The museum's director wasn't interested in birds, and so they were sold to an antiques dealer in 1957. Folks at U.S. Fish and Wildlife heard of the transaction and insisted that the dealer return the birds to the museum. A graduate student identified the Labrador Duck and convinced the people at the University of Vermont to donate the specimen to the outstanding museum at the University of Michigan. Storer catalogued the stuffed duck on May 21 as specimen number 152,253. Storer explained that there were no data to go with the specimen, but suspected it to be one of a pair of Labrador Ducks described on old lists as being in the collection of the University of Vermont. A search was made for the female, but she was never found.

So there you have it. There never was a garage sale duck. After searching for it for two years, I was able to dig up a photograph of the specimen in my files less than a yard from my left elbow, and Lisa and I were off to Ann Arbor to see it.

AT THIS POINT, I would have traded any two of my teeth for a good long sleep, preferably one that lasted until my next birthday. And yet, just ten hours after setting down in Calgary after the long flight from New Brunswick, I was in the air again, this time with Lisa. Having flown the width of North America the night before, I was now jetting the 1,364 miles back to Illinois. My duck quest was turning into an obsession, as quests so often do.

The first target of this quest was the 121st annual conference of the American Ornithologists' Union at the University of Illinois in Champaign-Urbana. I was scheduled to make a presentation on my birdsong research and chair a session. By delightful coincidence, the conference was not far from a couple of museums that housed Labrador Ducks. My duck business would follow the conference, but getting everything done would require rather precise timing. Lisa had booked passage on the 6:10 a.m. Amtrak train out of C-U for Chicago, where we would catch a taxi to the Field Museum of Natural History before returning to the train station at 2 p.m. for the journey east, bringing us into Ann Arbor, Michigan, around suppertime.

Monday morning we were up at 4:30 so as to get a hearty breakfast at the twenty-four-hour "We Never Close" restaurant next to our hotel. They were closed. "Sorry for the inconvenience" read the hand-printed sign on the door. Not an auspicious start. As instructed, we got to the train station twenty minutes before departure, only to be told that, due to mechanical difficulties, our train was running two hours late. The purveyor of this information had the look of someone for whom 90 percent of his job is passing along bad news to irate customers. I started to see why Amtrak is so unpopular with my American colleagues.

Incredibly, the train from the south spent the last fifteen minutes of its journey traveling backward into Chicago's Union Station at glacial speed. America's Second City. The Windy City. The City of Big Shoulders. Pride of the Rustbelt. That Toddling Town. Hog Butcher to the World. Somehow, backing into town several hours behind schedule didn't seem to be a fitting way to enter Chicago for the first time. Luckily, Lisa was able to convince an Amtrak representative to trade our 2 p.m. train tickets to Ann Arbor for tickets on the 6 p.m. train, giving us time to complete our Labrador Duck work.

When we arrived at Chicago's great Field Museum of Natural History, David Willard came straight down to meet us. Willard is the manager of the bird collection in the museum's zoology department. He was tall and clean-shaven, and sported the big black-rimmed glasses that are so hard to find in a world full of miserly wire-rimmed granny glasses. Willard grabbed one of our suitcases and we followed him up a flight of stairs to the museum's behind-the-scenes collections. Given our arrival time, he had guessed that something had gone wrong with the train. Not much of a stretch of imagination, he suggested.

Labrador Ducks 14 and 15

The ducks were waiting for me, and my examination of them was straightforward. Both are taxidermic mounts. The drake has a particularly jaunty look about him. He sits on a plastic rock atop an oval board. His face is more appealing on the left side, but his bill is prettier on the right side. The hen is a little more beaten up, missing a toenail, and with holes in the webbing of her feet. Her bill was painted with unflattering gobs of yellow and gray paint, now flaking off. Unlike the drake with his cool plastic rock, she resides on a simple wooden base. Except when nosey ornithologists are poking around, he has a plastic dust cover; she doesn't. His catalogue number is 13352 and hers is 13353.

With Willard's help, I had tried to sort out the origins of these ducks. It is abundantly clear that their origin is anything but abundantly clear. The Field Museum's database claims that both specimens came from the Charles B. Cory Collection after being shot at Grand Manan Island, New Brunswick, where Sarah and I had been a few days before. However, just because someone enters it in a computer database doesn't guarantee that the information is correct. Writing in 1959, Jim Baille of the Royal Ontario Museum claimed that the hen came from Swampscott, Massachusetts, about 1862, and that the drake came from Grand Manan Island about 1860. A note on file, in the handwriting of ornithologist Emmet Blake, explains that both specimens were shot near Calais, Maine, on dates unknown, and that the museum purchased them from George N. Boardman of Calais around 1880. The Calais locality seems to have been Boardman's home, but whether the ducks were shot there is anyone's guess. So, to summarize, the Labrador Ducks at the Field Museum in Chicago may have come from New Brunswick, Maine, Massachusetts, or somewhere else.

Having finished my work, and with some time to kill before our evening train, we told Willard that we were going to have a look at the museum's exhibits. He replied that we were very welcome to take in the exhibits, but as guests of the museum, we were not to pay the entry fee. God bless this man. We carried our luggage to the coat and baggage check, and Willard explained to the lady attendant that we were valued guests of the museum, and that she wasn't to charge us for watching our gear.

As we had seen, the two real Labrador Ducks are stored safely in locked cabinets, far from damage by light and dust. However, in tribute to creatures now extinct, the Field Museum has a replica duck on display, along with a replica Great Auk, and one real specimen each of extinct Carolina Parakeet, Passenger Pigeon, and Eskimo Curlew. As Lisa and I stood in front of this exhibit, a family walked up—parents accompanied by a son and daughter, both in their teens, and both with neon-blue hair. To my absolute delight, the girl looked at the iridescent yellow and red Carolina Parakeet and said, "Can you imagine being in a whole flock of those? That would be awesome!" I wanted to shake her hand for resisting the temptation to fall into the cynicism of youth, and for having the imagination to see that an extinction had denied her generation of a unique treasure.

We wandered through the other bird and mammals exhibitions and, like those patrons around us who had paid to be there, were very impressed. Stuffed specimens of fully 90 percent of all North American birds were on display, and each had sufficient room so as not to appear crowded. Some very large display cabinets had as few as one large horned mammal each.

Well into the afternoon, we stopped at the cafeteria for our first meal of the day. Shrink-wrapped white-bread sandwiches weren't good enough for the Field Museum cafeteria. Lisa had chili, and I had pizza on focaccia bread. As we ate, we had a good gawk at the architecture. The building was clearly designed to say: "The contents are important, and we want you to be impressed."

We missed the museum's exhibition on baseball, but we did see some of the other very good exhibits, including displays of gems and human history. Surely no one would fail to be completely impressed by one of the Museum's crowning glories—an almost compete skeleton of "Sue," a *Tyrannosaurus rex* discovered in North Dakota in 1990, and the biggest one ever discovered. Nothing about the film *Jurassic Park*, even on a large screen, can prepare you for the enormity of the real thing, and camera flashes popped all around us. A recently published paper has suggested that *T. rex* might have been a scavenger rather than a fierce predator, but part of me wants to think that these devils once terrorized the Cretaceous landscape.

The train pulled out of Chicago for Ann Arbor on time, but that was little comfort, because for the first hour, our speed didn't exceed 20 miles an hour. The journey was punctuated by long periods of immobility while the train caught its breath. Along with everyone else in our car, we tried to find seats as far away as possible from an obnoxious, loud, dysfunctional family. I realize that Amtrak is not responsible for the antisocial behavior of some of its clients, nor for the unappetizing view around the south shore of Lake Michigan. Where we should have been looking out over a Great Lake, we could see only factories, warehouses, steel mills, and water treatment plants.

How often can you say that the most beautiful thing on the horizon is a casino? This four-hour trip took six hours, getting us to Ann Arbor many hours behind schedule, long after dark, and way too late to get anything to eat. I swore to God Almighty that I would never get on an Amtrak train again.

Lisa never lost her good humor, but I needed something to give me a sense of perspective. Our cab driver provided just that. Farhan had arrived in Ann Arbor just two months before, having escaped from the war in Somalia. He was bright and happy and made me realize that, while I had the right to feel tired, I had no right to be grumpy. He offered to be our driver the next day, but we explained that we hoped to see Ann Arbor on foot.

I WAS NOT well rested the next morning, due in part to the inability of our very costly hotel to meet the needs of someone in need of sleep and a shower. Someone should explain to them that you can't turn a smoking room into a non-smoking room by simply taking out the ashtray when the guests arrive. When the pillows get to be as thin as the towels, it is probably time to toss them out. When the towels become as rough as the non-slip shower mat, it is probably time to replace those as well. I remembered Farhan, and the death and destruction in Somalia, and felt much less critical.

Whichever advertising genius described our hotel as "close" to the University of Michigan was probably the same person who described mad cow disease as a cure for high beef prices. At least we got to see a good chunk of the town as we walked north to the university to see my next duck. According to its promotional literature, the University of Michigan has more than 38,000 students. The institution brags that its graduates include one U.S. president, seven NASA astronauts, three Supreme Court justices, and 248 people convicted of illegal duplication and distribution of videotapes. The city has a population of 109,000, but Michigan Stadium seats 105,000 for university football games. Oddly, attendance at a single football game almost exactly matches the number of visitors to the University of Michigan's Museum of Art in a year. The art museum is free.

The streets around the university were lined with quirky cafés, second-hand bookstores, cheap bars, and clothing stores catering to fashionable college students. These establishments must all suffer quite badly when school is not in session. The university buildings themselves are opulent, and we had the impression that the architects had been influenced by the European tradition of post-secondary institutions, with lots of stonework and plenty of unnecessary ornamentation. This place is here to stay.

We found the Museum of Zoology but then walked around looking a little helpless until a hot dog vendor pointed out the front door. Admission to the museum's public displays is free, so we wandered in. The problem with this arrangement is that there is no reception desk and no one to direct us to the research collection. We spotted a door labeled Museum Personnel Only, found it to be unlocked, and strolled through. Up some stairs, down a corridor, around a corner, up an elevator, along another corridor, until we found Janet Hinshaw, Collections Manager of the Bird Division, working away at a desk in a back room.

Labrador Duck 16

I don't think that Hinshaw immediately registered who we were or what we were doing there. To give her credit, I had made the appointment to see the duck four months earlier. She recovered quickly and led Lisa and me to Interior Steel Equipment cabinet 27A, and brought out the garage sale duck. Hinshaw carried the duck to a workbench with good natural lighting, and I settled in. In most cases, curatorial types get on with their work and leave me to mine, but Hinshaw was pleased to chat with Lisa as I measured. We swapped a few stories as often as I could break my concentration without breaking the duck, and we had a pleasant little time.

The drake is a pleasing presentation, a taxidermic mount on a simple wooden base, stained and varnished. The feathers around his bill are a bit grease-stained, but at least no one had ruined his bill or feet by painting them. The duck's wooden base was inscribed "AJ Allbee and Sons," and an illegible name of a town in Vermont. Might Allbee have been the duck's taxidermist? Also visible on the base, below the name, were the words *Manufacturers of Doors, Sash, Blinds.* According to the website of the township of Derby in Vermont, A.J. Allbee operated a sash, door, and blind factory in Derby village in the nineteenth century, employing six hands and generating $5,000 worth of stock annually. Rather than having been put together by a taxidermist named Allbee, I think that someone made a base for the Labrador Duck using a discarded crate from the factory owned by Allbee. The duck probably has nothing to do with either Allbee or Vermont.

After finishing my work, Hinshaw gave Lisa and me a little tour. The University of Michigan's Museum of Zoology in Ann Arbor has about 200,000 stuffed bird specimens, which is 1.83 birds for every man, woman, and child in the city. Among these is an impressive collection of Kirtland's Warblers (on the road to recovery after a close brush with extinction) and Dusky Seaside Sparrows (gone after a full-on collision with extinction). Hinshaw seemed very proud of the scientific value of the specimens, including a stuffed extinct Heath Hen and the extinct-or-very-nearly-so Eskimo Curlew. Unlike other institutions of this sort, the museum wasn't smelly. The story goes that a long-gone curator had been allergic to mothballs, and instituted a policy of killing pests using a less fragrant insecticide.

And so our week-long duck adventure in America came to an end. In the morning we caught a cab to the Detroit Metropolitan Wayne County airport, and our plane took off just five minutes before a colossal blackout that blanketed most of eastern Canada and the northeastern United States and stranded airline passengers for several days.

Chapter Ten

Star-Crossed Ducks

Life is full of opportunities that seem like a good idea at the time. Time-share condominiums, lap dogs, and amateur dramatics come to mind. When I first received notice about a conference on European museum bird collections in Leiden, the Netherlands, it seemed like a perfect opportunity. I knew there were two stuffed Labrador Ducks in Leiden. The conference would be full of curators of European museums that I still needed to travel to. And the conference was a full year away.

The best way to be involved fully at a conference is to make a presentation. This provided something of a problem, since I was likely to be the only person at the conference who was not involved in the curatorial care of stuffed birds in Europe. As an outsider, might a useful (perhaps "irritating") topic be a summary of all of the things I thought curators were doing wrong? A user's perspective on museum cock-ups, as it were?

Not quite brave enough to venture into this myself, I contacted Errol Fuller, who had made his feelings about the cock-ups of museum curators very clear on many occasions. Errol indicated that he would be pleased to co-author a presented paper in Leiden and take half of the grief for it. So we submitted an abstract, which was immediately accepted by the conference organizers, and wrote the date on our calendars. After discussing topics we thought most important, I put together a supporting presentation, booked my transatlantic flights, reserved a hotel room in Leiden, and paid my conference fees.

Oh, but surely the conference would not be enough to fill a weekend. Sixteen time zones, seemingly unending flights, a conference, and an oral presentation—surely I could do better than that. Remembering a colleague at the university in Leiden, Hans Slabbekoorn, who also did research on the songs of birds, I offered to give a presentation on that bit of my research just before the conference started. Done!

Luckily, the conference happened to fall during the Thanksgiving long weekend in Canada. So I could, by missing just three lectures, fly across the Atlantic, ignore jet lag brought on by the first eight time zones, spend about ninety minutes measuring the two Labrador Ducks at the national museum, make a sixty-minute presentation at the university, rip back to make a twenty-minute presentation at the conference, meet a lot of people and shake a lot of hands, eat some Dutch pancakes, drink as much foreign beer as I could, and then zip back across another eight time zones, just in time to give my Tuesday lecture at 8 a.m.

When reality set in, I had to give serious consideration to my loss of sanity. It was costing me several thousand dollars for flights, $400 for a hotel room, and $300 for conference fees, and the only tangible result would be my written notes on the colors and linear dimensions of two more Labrador Ducks. But just to make it seem a little worse, the day before flying out of Calgary, I received an email message from Errol explaining that some unspeakable horror had come up, and that he might not be able to join me at the conference. He hoped to jump on a flight at the last minute, but I wasn't to hold my breath.

I WOKE TO face the morning of an action-packed Friday in Leiden with a headache. It has always seemed unfair to me to wake to a hangover not preceded by heavy drinking. Suspecting dehydration, I downed a lot of juice and milk with my complimentary breakfast in the hotel lounge.

Setting off for the museum, which I knew to be quite close to the train station, I clutched the city map that I had printed from the

Internet. Having walked about two blocks, I cursed my forty-five-year-old eyes and my long day of travel, because I couldn't make out any of the tiny street names on the map. And so I resorted to my time-honored mode of orientation in a strange city—I followed the main flow of traffic on the assumption that everyone must be going to the same place I was.

Vehicular traffic was light, but there was a steady stream of bicycles, all propelled with a great sense of purpose. These were not the high-tech mountain-bike–racing bike hybrids I was accustomed to, but something that I remember from the 1960s, with wide fenders, a broad and sensible seat, and a small satchel mounted under the seat with a single wrench for roadside repair. It'll get you there and back and, best of all, no one would want to steal it. Among cyclists, schoolchildren outnumbered shopkeepers and office workers by about two to one.

Going in the right direction or not, I had a wonderful ramble. I was the only person at that hour without a coat on, something to do with being a cold-hardened Canadian, I suppose. The tide of commuters took me through tidy mid- and high-density housing, three and four stories high. With sidewalks, bicycle corridors, and automobile lanes, all neatly partitioned off, Leiden seemed a sensible and orderly place. By the time the morning commute had swept me to the train station, I could see the gleaming tower of Naturalis, just a few blocks away, beckoning me. Once again, my version of Zen navigation had served me well.

Naturalis is the Dutch National Museum of Natural History, founded in 1820. The first King Willem assembled it by drawing together several large collections of natural history artifacts into a single museum. Early in the twentieth century the collection was moved to a specifically designed museum building in Leiden. That wasn't sufficient for the good people of the Netherlands, however, and they constructed a sparkling new facility, opening to the public in 1998. As evidence of just how wonderful this facility is, in 2001 the turnstile admitted its one millionth visitor.

This is no little backwater penny-ante research facility. The collection includes more than 5 million insects, 2 million other spineless animals, and 570,000 braver animals with backbones. For those who like their natural history to be really long dead, there are 1,160,000 fossils, as well as 440,000 stones and minerals. Among recently extinct species in the collection are a quagga (think of a cross between a horse and a zebra), a thylacine (Tasmanian marsupial wolf), a blaauwbok (something that a cat hacked up), both Cape and Barbary lions, a Javan Lapwing, a Great Auk, and two Labrador Ducks. If you desperately want to see a stuffed White-winged Sandpiper, a trip to Leiden is in your future, because Naturalis has the only one.

A portion of this amazing collection resides in the part of the museum open to the public. I was scheduled to see that a bit later in the weekend. From the outside it looked cheery and welcoming. The vast majority of the research collection is housed separately in a great silver monolith towering 200 feet above the city.

The collection building is awe-inspiring. In a city where most buildings are just three or four stories, this tower was impossible to miss. Not a window perforated its exterior surface, which is sculpted like the skin of a snake. Security was tight. My appointment was for 8:30, and in typical fashion I arrived early. I was permitted through a hermetically sealed door and invited to sit in an alcove where I could neither advance nor retreat; in essence, I could cause no trouble whatsoever. René Dekker, Curator of Birds at Naturalis, had warned me that he might be a bit late because of traffic congestion, and so I settled in for a little wait. As other museum employees arrived, each offered me cheerful greetings, and although I couldn't quite make out what they were saying, no one seemed to be insulted by my reply of "Good morning."

Eventually, Dekker swept in. He seemed exactly the sort of fellow who should be featured in television advertisements for sports cars—enthusiastic, friendly, and accommodating, with a glint in his eyes that said he would never grow old. We flicked through one security door after another; it was apparent that if you were to somehow sneak into the building, you wouldn't get far. Or out. Dekker settled me in at a work desk and locked the door behind him, and I got down to examining my ducks.

Labrador Ducks 17 and 18

The drake and hen pair in the Dutch National Museum of Natural History in Leiden have resided side by side since 1863.

Both the hen and the drake were in really fine shape. The feet were unpainted, and not nearly so bashed up as most specimens I had seen. The taxidermist had been judicious in painting the bills of both specimens, with very little paint splashed up on feathers above the bills. The right side of the male's face around the eye and cheek had seen better days; perhaps that was the side where he had been shot. The female had patches of glaucous feathers, particularly on her wings, which I had not encountered on earlier specimens. Both birds stood on small wooden bases, painted white. In both cases, in neat script on a card in black ink above carefully penciled guide lines, were the words *Voyage du Prince de Neuwied. Acquis en 1863 Amérique du Nord.*

The slightly longer version of the story says German naturalist Prinz (Prince) Maximilian zu Wied-Neuwied collected the ducks when he traveled to North America early in the nineteenth century. Museum officials in Leiden acquired the pair in 1863, probably as part of an exchange of specimens with Vienna's Naturhistorisches Museum. It's anyone's guess whether Prince Neuwied ever visited the regions where Labrador Ducks were found and shot them himself, or traded for them, or found them dead at the side of a busy carriageway. Today, when nosy ornithologists aren't poking and peering at them, the ducks sit side by side in a special room for the museum's most valued specimens, with dim lighting and heavy security.

It may have been jet lag, and it may have been a lack of sleep, but a very curious feeling came over me as I examined this pair. As a couple, they seemed unlucky. They were due to spend some approximation of eternity side by side, but never really together. A good guess is that they will outlast me by about 450 years, but only as star-crossed lovers. Heaven knows, they may have been shot decades apart, or they may have been collected as a mated pair—Prince Neuwied didn't say. Perhaps they hatched a brood of cute little Labrador Ducklings in the nether regions of northern Canada and guided them to the wintering grounds before getting blasted by a collector. Clearly, I was getting a little lightheaded. After nearly two hours of measurements and careful consideration, I telephoned Dekker so that he could let me out of the locked room and escort me from the building.

LEIDEN HAS BEEN strongly influenced by its university, but unlike Champaign-Urbana or Ann Arbor, this influence has been going on for more than seven centuries. The center of the community is full of the sorts of restaurants, cafés, and bars that students and tourists alike enjoy. One large square featured a number of fast-food joints as well as three restaurants distinguished by the names 't Panne koekenhuysje, Pannekoeke Backer van Dam, and Pannekoekenhuis, all presumably featuring pancakes. I wandered past a windmill, along canals, and over small arched bridges. Surprisingly, there seemed to be no park benches along the canals, so I couldn't kick back for a few minutes to watch the citizens of Leiden flow by me.

I found the Evolutionary Biology Building, right where Slabbekoorn said it would be. He met me in the reception area, and I was slightly disappointed with myself to feel jealous that Slabbekoorn was more handsome than any scientist has the right to be. We chatted about life and birds and research in his office, and he then took me on a tour of the building, introducing me to professors, postdoctoral fellows, and graduate students. This single building had more people working in the field of cultural evolution than in all of Canada.

After lunch, a healthy group assembled in a lecture hall for my seminar, and I dug in. I had prepared a slide presentation rather than using PowerPoint. Good thing, too, as the computer projection unit had been stolen, ripped clean out of the ceiling. The topic was an overview of my work on songs of sparrows and New World warblers, and I was a little worried about getting across subtle concepts to a group whose first language was not English. I needn't have worried, though, as my version of English was better understood in Leiden than it had been in Dublin. Along with methodology, results, and interpretation, I tried to throw in some anecdotes about the wilds of Canada (grizzly bear stories always go over well in Europe), and laughs came at the appropriate times. My seminar was followed by another ninety minutes of talks and tours with gracious folks in the department.

I retraced my steps through the city, over canal bridges and along centuries-old streets, back to Naturalis for the start of the conference, to find the opening reception in full swing. I quickly found Clem Fisher from Liverpool, and a group from the Natural History Museum in Tring, most of whom I had met on previous Labrador Duck quests. We spent the next hour drinking Dutch beer. I found myself beside Katrina Cook, who had just begun an appointment as a curator at Tring. She had the dark, sultry good looks and sexy swagger of an Eastern European who might have seduced British spies in the Cold War era. With bellies full of beer, we listened to an opening address by Michael Walters (Santa Claus) who told us about his life as curator of the world's largest collection of bird eggs at Tring. We then sallied forth into the night in search of more beer and perhaps some food.

A very large group set off, but as we snaked through the early Friday evening, the group budded and then divided again. I was keen on pancakes, but the remainder of the group wanted Mediterranean food. I found myself in a Greek restaurant with a group of twelve, including Bob McGowan from Edinburgh, Fisher from Liverpool, and Katrina, Russell, Walters, and Prys-Jones from Tring. My two highest priorities were an exchange of lofty thoughts and getting something into my stomach that would sop up the beer, and so I ordered the first vegetarian dish on the menu. And more beer.

We drank and laughed and drank some more. Food was mixed

in there somewhere. The restaurant was typically European, fitting more patrons in a small space than my high school geometry classes said should have been possible. I was the filling in a sandwich between Katrina and Fisher. Katrina proved herself very good at flirting. Rather than ask me to pass her the salt, she put her right arm around my shoulders and reached with her left. Her leg grazed mine more often than expected by chance. She spoke to me in a breathy voice that grazed my cheek. Later that evening, while walking back to her hotel, she fell into a canal while peeing behind a bush. Sneaky things, those Dutch canals.

RIGHT UP TO the last moment, I had expected Errol to show up to take his share of the blame for our talk. There was no sign of him. I tried not to see it as an omen that the building I was to present in, the Pesthuis, had been constructed in the 1650s to house victims of bubonic plague. Surrounding the facility was a moat, although it was officially described as a canal. We learned that there had been a second moat separating the men on one half of the compound from women on the other half. Call me a hopeless romantic, or call me an incurable sex maniac, but if I knew that I had only twenty-four hours to live, and was separated from a willing partner by a moat, I think that I could find the energy for a quick swim. After the plague epidemic passed, the facility was used as a prison for military convicts, and then as a mental hospital.

In a move almost guaranteed to inflame, my talk was scheduled as the first one of the conference. I was nervous. After all, I was about to give the group a long list of the things that they were doing wrong. Well, not exactly, I suppose. I was going to give them a long list of things that could be done differently to make the life of the museum user a bit easier. Each conference attendee had received the abstract of each talk in advance, and perhaps I was being a little paranoid, but I got a sense that the audience might have been a bit tense, like spectators at a public execution.

I launched into the twenty-five-minute talk, entitled Views from the Other Side of the Bird: Users' Perspectives on Bird Collections. It came off well enough. I didn't stumble too often, although I was more reliant on my prepared notes than usual. A few chuckles came at appropriate times. When my allotted time was up, a little yellow light illuminated on a panel on the podium. It was beside an equally small sign saying *Stopen.* It seemed a lot more friendly than the German equivalent, *Halten,* and so I politely stopped speaking, and invited questions and comments. There was a stony silence. Dekker stepped in and lobbed me an easy question about the use of museum collections by captive bird breeders, which allowed me to emphasize a couple of my earlier points. This opened the floodgates, and questions and comments came flying. One fellow took particular exception to my suggestion that policies about collection use should be explicit and published, claiming that museums should have the flexibility to make decisions on a case-by-case basis. I countered that explicit policies helped to ensure that decisions were neither arbitrary nor petty. Luckily, I had already done all the work I needed to at his institution, because I think it unlikely that I will ever see the inside of it again. Afterward, a junior member of the same museum took me aside and in a small voice claimed that my talk had "said what needed to be said."

The remainder of the day was filled with talks and a formal tour of the museum, including the research collection. The tour of the gleaming new facility was preceded by a talk by the building's architect, who was clearly proud of the climate-controlled closed system that he had designed. The interior of the building never wavers from 67 degrees Fahrenheit, and the relative humidity is always exactly 54 percent. It was equally clear that many of the curators from other institutions were envious, if only of the endless row of cabinets that were not yet filled to overflowing. I gather that space is at a premium in most facilities.

We were shown one room after another full of stuffed birds, some as study skins and some as taxidermic mounts. I pulled out a tray with an impressive series of birds from New Guinea called pitohuis. I called over one of my Spanish colleagues and asked if he could do me a favor. "Sure," he said. I asked, "Would you mind licking one of these birds for me?" "Sure . . . what? . . . why?" I explained that it had been discovered recently that, as a means to defend themselves against predators, pitohuis had evolved poisonous feathers. It is said that if you lick a pitohui, your tongue goes numb. Or you might die.

I wanted to know if it were true. He gave me a disgusted look and walked away. Perhaps my humor isn't sufficiently subtle.

We were then given the opportunity to tour the myriad exhibits on public display. Everywhere there was something interesting to see: stuffed birds and mammals, skeletons of whales and dinosaurs, insects, shells and fossils; all were displayed to advantage. Cooperative signs offered explanation to visitors in Dutch, English, and French. In a cabinet constructed of bulletproof glass was the skullcap, molar, and leg bone of a near-human creature that lived about 1 million years ago, collected on the island of Java in 1891. I confess that I didn't have the energy to do justice to the assembled menagerie. It is the sort of facility that has to be approached fresh, with the energy and enthusiasm of a child.

Afterwards, catching a second breath, the conference mob again hit the streets of Leiden. We passed only two bars before settling on one situated beside a canal, advertising Heineken in large, friendly letters. I downed one Heineken before settling in for a good long sup on wheat beer. I didn't get the chance to pay until the fourth round, by which time the bull was really flying. There was a lot of talk about colleagues who weren't in attendance. Someone suggested that it was really mean of Errol to leave me high and dry, but someone else suggested that his absence greatly decreased the probability of a punch-up. When challenged, Fisher swore that she was not the granddaughter of the Archbishop of Canterbury, with the sort of energy that suggested that she was. As tipsy as any of us, she said, "Would the granddaughter of the Archbishop of Canterbury use the expression 'Up yours'?"

WALKING FROM THE hotel to Naturalis for the final day of presentations, I had a bit of fun with street signs and billboards. I speculated that a building, labeled Leids Universitair Medisch Centrum, was a place where I could be treated for plague, should the need arise. "Te koop" seemed to indicate that an empty home was for sale, or perhaps awaiting demolition. The sign at a construction site proclaiming "Onderzoeksgebouw LUMC" completely befuddled me. On a quiet residential street, I cobbled together contextual clues, and gathered that "aanbiedplaats minicontainers niet parkeren UO 7.00–16.00

UUR" meant that between 7 a.m. and 4 p.m., I was not to park my minivan in front of an aanbiedplaats.

And so the day was occupied by talks about type specimens, about the collections of specific museums, and about the use of museum specimens as sources of DNA. Some of it seemed a bit much like stamp collecting, but with birds instead of stamps. I was pleased that several people made reference to the presentation I had made the day before, and that none of those people used vulgar language. After the last talk, we all gathered for a group photograph. There was a brief discussion about publishing the proceedings of the conference, and about the next European Conference on Bird Collections, to be held two years hence in Spain. With that, the Leiden conference came to an end.

At most conferences I had been to, the last evening was occupied by a banquet, and attendees dispersed back to their home countries the next day. In contrast, almost all of the people in Leiden were driving or flying away on Sunday evening. I wasn't scheduled to fly out of Amsterdam until the following morning. As much as I had enjoyed Leiden, I didn't particularly relish the thought of spending my last evening in it alone. Luckily, Katrina was staying over, as she had some work to do at Naturalis the next day. We agreed to make an evening of it.

We went in search of beer and food. Since Katrina claimed no special preference, I chose a pancake restaurant. I ordered a pancake with cheese and mushrooms, and Katrina opted for something with spicy beef. There was a mix-up with the order, and Katrina got cheese and ham but decided to eat it anyway. She was talking intently, and I was listening intently, and after a few mouthfuls, it was apparent that there was a second mistake with our order. I had been eating the spicy beef pancake. To help explain myself, after twenty-five years as a vegetarian, I really don't remember what meat tastes like.

We walked along Leiden's canal system, gabbing about loves won and lost. We walked past several grimy-looking bars, but eventually found a good one that was broadcasting gentle jazz workings into the night. The Dutch have no concept of a large glass of beer, and so we had to settle for beverages in little glasses, one after another. We talked about art and birds and other bits of nature. We flitted between topics like global climate change and the latest theories on human sexuality.

And then it occurred to me that Katrina had been carrying the few belongings she needed for the three-day conference with her. With the end of the conference, she had lost her roommate and was now in need of a hotel room for the night. It seemed to me that her chances of getting a reasonably priced hotel room at 10:00 on a Sunday evening in an unfamiliar city in a foreign country were rapidly diminishing to zero. Putting the offer in as gentlemanly a way as possible, I offered her the second bed in my room. She accepted, and I got another round of beer.

Rather than paying a few extra euros, and registering her as an official hotel guest in my room, Katrina preferred to sneak past the front desk as I was retrieving my room key. Perhaps she liked a sense of danger. Perhaps she was really short on cash. While she was in the bathroom, getting ready for bed, I changed into the green cotton surgical garb that I use as pajamas when the situation requires me to wear clothes to bed. These apparently disturbed Katrina, so I switched into a gray t-shirt.

At one point in the night, Katrina woke a bit, and in the pitch dark room called out, "Are you still there?" Trying to call myself from a deep sleep, it came across as "Bzz bzz bzzzz bzzzz?" She called out again, a little louder, "Are you still there?" "Yes," I said, "I'm still here. Go back to sleep."

Having not paid for her part of the room, Katrina was not keen to join me in the hotel breakfast foyer for breakfast. I lingered over toast, scrambled eggs, and juice, and in an attempt to keep up appearances as a gentleman, I brought coffee and a croissant back to the room. She snuck by the reception desk as I checked out. We walked into town so that I could catch the train to Amsterdam's airport and she could get a proper breakfast before the museum opened. At the train station, at the top of the stairs to my platform, I looked back. Katrina was waiting to wave good-bye.

Two more ducks were behind me, but I felt it was time to quicken my pace. I had two semesters of teaching ahead of me, but then my university had granted me a year-long research sabbatical, and that would be enough to snap up every remaining duck in the world. If everything went according to plan.

Chapter Eleven

Enduring Images of Germany

Altenburg is a tiny community in the northeast corner of Germany, not far south of Leipzig and an equally small distance west of Chemnitz. Chances are you have never heard of Altenburg and will never hear of it ever again. From all of the material sent to me by the Altenburg tourist information bureau, it seems a perfectly charming place. The most exciting thing that ever happened in Altenburg was the kidnapping of a couple of young princes by the evil warrior knight Kunz von Kaufungen in 1455. The princes were retrieved none the worse for wear, and naughty von Kaufungen was publicly beheaded in the town square. The second-most-exciting thing to happen in Altenburg was the development of the card game skat between 1810 and 1815. Although I have never played skat, I am sure that it is big barrels of fun—the sort of thing you might turn to if German television ever gets rid of free porn.

Altenburg's Naturkundliches Museum Mauritianum has a modest collection of dead birds. When Paul Hahn wrote to the museum in the late 1950s, asking about stuffed specimens of extinct North American birds, H. Grosse, the museum's director, wrote back to say that he was in charge of one Eskimo Curlew, one Ivory-billed Woodpecker, two Passenger Pigeons, one Whooping Crane, one Carolina

Parakeet, and, best of the lot, one Labrador Duck. Grosse didn't provide a long treatise on where, when, or by whom the duck was collected, or how the museum came to have it. Instead, he wrote "one." This makes it the single most enigmatic Labrador Duck specimen in the world, with no information about its sex or age, or whether it ever really existed.

The current science director of the Altenburg museum, Dr. Norbert Höser, told me that he has examined the collection and found no Labrador Duck specimens. And so, despite the opportunity to visit a community famous for the manufacture of playing cards for more than four hundred years, with a museum of playing cards to prove it, I did not travel to Altenburg, Germany. (If you do, and see a stuffed Labrador Duck, it will be worth your while to read this book's epilogue.) I did, however, have seven other German cities to visit.

THERE WAS A time when airline travel was something a bit elegant. I've seen the photographs. Men wore business suits and smart hats, and women wore elegant dresses. In the era when only those with quite a bit of spare cash could afford to fly, I am sure that in-flight emergencies were greeted with expressions like "What's that? An imminent crash, you say? Oh dear. Bad show! One had better buckle up, I suppose." Today, any old rabble can fly. Me, for instance, and I find myself asking if I want to fly with any airline that would have me as a passenger.

But there I was, buckled into a seat on an ultra-low-cost flight from Glasgow to Frankfurt. These unbelievably inexpensive airlines are able to cut costs by using some of the less popular airports, including facilities normally used by crop dusters and zeppelins. In order to save myself a few pounds, flying into "Frankfurt" didn't mean arriving at the great hub of international transportation, and home to a wide selection of cafés and bars and Germany's only airport sex shop and X-rated movie theater. Instead, I arrived at an airport adjacent to the minuscule community of Hahn, in an entirely different state. The airline must save even more by flying at unpopular hours, and so my cheap flight to "Frankfurt" deposited me several hours from Frankfurt, a few minutes before midnight. After explaining to a very polite but incredulous customs agent that I was in his country to see dead ducks, I went in search of my hotel with a website claiming that it was just 800 meters from the terminal. This statistic is known in the business world as "creative advertising."

The next morning, my luggage and I wandered "800 meters" back to the airport to wait for the bus into Frankfurt. I passed what appeared to be an American military base. A big billboard for an American sports bar described itself as "Your favorite place on the base." With no frame of reference, I couldn't disagree. A couple of flights disgorged their passengers just before the coach departed, and there wasn't room for everyone. As we pulled away from the terminal, I saw a dozen travel-weary and dispirited passengers left behind on the sidewalk.

The bus carried us past grain fields, soon replaced by deciduous forests perforated by massive wind turbine farms, a few orchards, and a very few vineyards. We dropped folks off at the real Frankfurt airport and then proceeded to the *Hauptbahnhof*, from which I needed to catch several trains over the next few days. To get a sense of orientation, I wandered through the train station and picked up a sandwich for €2.60. I mention this only because I like to use the "€" button on the keyboard. I used the WC (€0.70), and stored my luggage in a coin-operated locker (€3.00).

It was too early in the day to go to my hotel, and too early in the week to see the Frankfurt duck, so I set off in search of adventure. At first it rained lightly, which was surely no impediment for a jaunty traveler. Then it rained more heavily, which was nothing to an intrepid adventurer, unless he had left his umbrella at home and his rain hat in his luggage in a train station locker. I walked east through the heart of Frankfurt's financial district. I saw a lot of gleaming great office towers, top-end hi-fi stores, and men and women in tidy haircuts and expensive gray suits. I also saw the most amazing array of sex movie houses, continuous live sex shows, and shops for women's underwear of the sort that aren't really meant to be worn under anything at all. The passing traffic seemed to be mainly composed of Mercedeses, BMWs, and Audis, with a few Ferraris thrown in to liven up the mix. There was also almost endless opportunity to drink beer at sidewalk cafés, had the weather been better, and if the neighborhood hadn't been full of so many scary, nonbanking characters. I take it that this part of Frankfurt is the German center for banking, prostitution, and nonprescription drug abuse.

Knowing that I couldn't possibly get any wetter, I continued my walk east and then cut south to the river, crossing to the far side at the Alte Brücke bridge. The Main is a truly lovely river, although almost completely devoid of commercial or pleasure craft. One entirely beautiful exception was a luxury cruise ship, registered at Strasbourg, tied up near the Untermainbrücke. Despite being absolutely spotless, it was being scrubbed vigorously by some members of the crew, while others laid a sumptuous meal for guests. This is the sort of ship that I will only ever see from the outside.

As a banking giant, Frankfurt has its share of very tall buildings. Most are simple, black, imposing, and uninspiring, which may be exactly the sort of image that banks wish to project. Offices of the DZ Bank reside in an uninspired silver-and-glass tower that is saved by a whimsical statue of a giant necktie at the base. Some of Frankfurt's skyscrapers are designed in a style that I think of as seemed-like-a-good-idea-at-the-time. For instance, the Messeturm is an 840-foot-tall building that starts off at ground level as a box. As it rises, its cross-section turns from a square to a star with too many points. Further up, the building loses all of its points and becomes a column. Topping the column is a pyramid. This has led Frankfurt residents to call the building *Bleistift*, but I swear it looks a lot more like a crayon than a pencil. Even the Marriott chain got into the act with an impressively tall building in which I will be invited to stay immediately after I get a ride on the Strasbourg luxury cruise ship. But aside from these exceptionally tall and grand towers, Frankfurt seems to have resisted the temptation to build upward, thereby ruining the view for everyone else.

After a shower and a quick nap at my hotel, I was off in search of more of Frankfurt. I was getting quite brave when it came to restaurant meals in foreign cities. Walking down Münchenerstrasse, I found a restaurant without white linen tablecloths, featuring both Indian and Italian dishes. Finding a column in the menu labeled *Vegetarische Gerichte*, I picked something with the word *milder* in the description. It was based on curried tofu and rice, and went very well with tonic water, which, strangely, translates into German as *tonic water.*

Bless their little hearts; the editors of the Lonely Planet guide to travel in Germany provide descriptions of "Dangers and Annoyances" for each city. The guide is quite clear that the region around the *Hauptbahnhof* is the navel of Frankfurt's underbelly. It goes on to explain attempts to tidy up the streets a bit. This involves *Druckräume*, buildings where an addict can inject drugs, away from prying eyes, and receive clean needles. I was even told where to find such a building, should I need one. Even so, the guide explained, I could expect to see junkies injecting and defecating on the streets near the main train station, and I wasn't let down.

THE NEXT MORNING I strolled to the Forschungsinstitut und Naturkundemuseum Senckenberg, in search of my first stuffed Labrador Duck of the trip. The museum is housed in a magnificent old building surrounded by Frankfurt University. I started my visit by sitting in the park that constitutes the median of Senckenberganlage, so that I could have a good look at the museum. Portions of the building are constructed of the most beautiful salmon-colored stones with tan veins. At the peak of the building, above the main entrance, is a statue of a naked old man with wings, holding a scythe in his right hand and an hourglass in his left. Flanking him are a couple of cherubs, with a couple of nymphets a bit further along. Beyond the nymphets, on the right, a man is riding a half fish, half horse. On the left is an ample woman riding a half fish, half cow. (Some sort of early trial in genetic engineering gone wrong, I suppose.) As I snapped a couple of photographs of the museum, great hordes of schoolchildren arrived. They were enthusiastic but well behaved, and their handlers subdivided them into small groups to keep them from reaching critical mass.

I tucked myself into their throng and entered the museum. The lady at the till was rather too busy taking in the admission fee of hundreds of students, and so I found a likely-looking helper near the turnstiles. "*Guten Morgen*," I began. "*Mein Name ist Professor Glen Chilton*." "*Ja*," he replied. "*Sprechen Sie English?*" I tried. And then in an accent straight out of Las Vegas, he said, "Sure, whadda ya looking for?" My new friend paged my contact, Gerald Mayr, a renowned researcher of avian fossils and the museum's curator of ornithology.

When Mayr entered the throng a few minutes later, Mr. Nevada gestured at him, so that I couldn't miss him. Like many other people in the business, Mayr is wildly enthusiastic about his work, and he was keen to make my experience entirely pleasant. He whisked me away from the crowd through secret doors to the behind-the-scenes world of the museum's research collections.

The Frankfurt museum collection was founded in the early nineteenth century by the Senckenbergische Naturforschende Gesellschaft, but it was based on the even older collection of B. Meyer. It contains about 90,000 bird specimens of 6,000 or 7,000 species. In addition, the collection has 4,000 skeletons, 3,375 birds in alcohol, 5,050 sets of eggs, and many fossils. Indeed, much of the scholarly work by museum personnel is done on bird fossils, or "paleornithology," if you are trying to impress your friends with your big vocabulary.

Labrador Duck 19

This Labrador Duck didn't come to rest in Frankfurt in the nineteenth century along with most of the rest of the collection. It was acquired in an exchange with the American Museum of Natural History in New York in December 1931. As rare and beautiful as Labrador Ducks are, Frankfurt got kind of cheated on the swap by New York's Dr. L. Sandford. Frankfurt received the Labrador Duck in exchange for a type of finch that lived on the Japanese Bonin Islands before becoming extinct. Stuffed Bonin Island Grosbeaks are even rarer than Labrador Ducks, but, having two, the folks at Frankfurt were willing to swap their female for a duck.

Mayr set me up in the *Bibliothek*, and brought me a tray with three birds, all extinct. The first was a Pink-headed Duck, the second was a Norfolk Island version of a pigeon from New Zealand, and the third was my Labrador Duck. Mayr's predecessor, Professor Dr. D. Stefan Peters, had sent me photographs of the specimen some years before, and so this one was a bit of an anticlimax. It is an immature bird, and although it is a study skin today, the wires running through its legs suggest that it might have once been a taxidermic preparation.

There cannot be much doubt that it was a male, as areas destined to be white were lightening up, and areas destined to be black had been getting darker before the terminal shotgun blast.

After taking my measurements, and putting the Labrador Duck safely back in his cabinet, Mayr showed me some of the museum's other treasures, including a stack of Carolina Parakeets and a selection of extinct Hawaiian birds; the bright feathers of these species had been used to construct ceremonial robes for Hawaiian kings, which probably hadn't helped their long-term survival prospects.

In a story that I was to hear again and again in Europe, Mayr explained that, during the war, Frankfurt's natural history collection had been split up and moved to several sites outside the city, which explained how so many specimens had avoided being blown to kingdom come during Allied bombing. We also chatted about the city of Frankfurt in general terms. He said that many first-time visitors were quite shocked by the overt use of hard drugs, and put it down, in part, to Frankfurt's liberal treatment of drug users. I said that I had been rather surprised by the overt prostitution and live sex shows. "Yes," he said, "but that's not Frankfurt. That's Germany!"

And so having completed my official work, Mayr took me back through the collection and released me into the public displays. Not for the last time, I got to see a great museum without paying for it. And this was truly a great museum, assembled with love and care, of the sort that I wish my university students had access to. There was a curious mixture of displays, both traditional and contemporary. The dioramas that featured large mammals were so well constructed and subtly lit that I was moved to whisper so as not to disturb the scenes. A little further along, stuffed birds, mammals, and fish were on display in supermodern glass cabinets with good lighting, so that if I were to set out to learn how to draw animals, this would probably be a very good place to start. For me, a highlight was the display of specimens of *Riesenalk*, the Great Auk. Museum visitors were able to see a stuffed bird, a skeleton, and an egg, although the last may have been a model. Also on display was a Dodo skeleton, and the skeletons of three species of extinct New Zealand moas. Display cards that accompanied many of the bird specimens indicated their conservation status. On the card in front of endangered specimens like the Kakapo was a big red circle. Three-quarters of the circle was filled in red for less-endangered species like the Sun Conure, and a red semicircle, like the one for the Purple-naped Lory, indicated that the species wasn't quite so entirely doomed yet. Accompanying extinct species like the Carolina Parakeet was a small map of the world with a big red X through it, indicating their current global distribution.

A very large gallery was devoted to dinosaurs. While examining the *Tyrannosaurus*, I had a delightful little experience. Three children, two boys and a girl, all about ten years old, came up to me and said, "*Entschuldigen Sie bitte*," followed by a lot of German that I couldn't keep up with. Instead of saying something vaguely German like "*Ich spreche kaum Deutsch*," or "*Ich verstehe nicht*," I said, "I'm sorry, but I don't speak German." They looked at me hopefully, and thrust a small disposable camera at me." "Oh, you want me to take your photograph?" "*Ja!*" "In front of the *Tyrannosaurus?*" "*Ja!*" I was pleased to find that society hadn't frightened them into avoiding contact with strangers at all cost. I crouched down as far as I could to frame them with as much of the dinosaur as possible, and clicked off a photo. "*Danke. Danke.*"

I set off in search of more Frankfurt sights. Tour books speak of the not-to-be-missed opportunity to scan all of Frankfurt afforded by the observation deck of the Main Tower. After paying the €4.50 entry fee, I had to pass through a security check as rigorous as that at any airport. I was directed to walk through a metal detector, while my keys and coins were thoroughly x-rayed and found to be harmless. A lady with a hand wand checked my trouser zipper with a degree of enthusiasm that didn't seem entirely professional.

After a high-speed elevator ride, I found the view from the observation deck to be truly grand, even though it reminded me that I suffer from a bit of vertigo. The tower's observation platform is 650 feet above the street, fully four times higher than the Arc de Triomphe in Paris. If I were a young lover calling Frankfurt home, I could imagine proposing marriage on the deck on a summer evening as the daylight faded. I could see that Frankfurt had only twelve or fifteen really tall structures, depending on the definition of "really tall," as most other buildings restricted themselves to just six or seven stories. I spotted the natural history museum with ease, as well as the Frankfurt airport in the distance. I could see the *Dom*, and the small Römerberg district that somehow missed being destroyed by Allied bombing. But then I had a disquieting thought. I suppose mine was much like the view of Frankfurt for Allied bomber pilots that had come to destroy the city.

With my feet back at street level, I aimed for the Römerberg, a district of cafés, trinket shops, bakeries, jewelry stores, and a rather peculiar toy store that featured in its windows an impressive assortment of railway cars, teddy bears, and playing cards with pictures of naked women. I didn't buy any. Honestly. I did buy a sandwich, chowing down while looking up at Frankfurt's *Dom*, which is not really a cathedral because it doesn't have an archbishop. Even so, it is quite a sight, constructed of the same sort of beautiful sandstone that was so prominent on the museum. At least, the chunks that I could see were constructed of sandstone—much of the church was hidden from view. Completion of renovation efforts were long overdue, but it was still being rebuilt when I visited in 2004, and the main tower was hidden in a multistory drape. Curiously, this shroud had a giant picture of a racecar advertising Panasonic. I had to wonder what God would think about having His house of worship covered in a giant advertisement. I hope that He would be pragmatic and not at all vengeful.

THE NEXT DAY I earned a few points for gall, at least in my own mind. The only reason I had for traveling to Mainz was a letter written in 1959, housed in the files of Paul Hahn at the Royal Ontario Museum. Someone with a scrawled signature had returned Hahn's letter of inquiry, indicating that the three Passenger Pigeons, Great Auk, and Labrador Duck in the collection of the Naturhistorisches Museum in Mainz had all been burned. No further details were provided. Presumably the fire had something to do with World War II. I had sent letters and email messages to curator and zoologist Ulrich Schmidt, but these had all gone unanswered. This could, of course, be taken as a signal to stop pestering him, but a small nagging voice inside me urged me to travel to Mainz anyway, just in case the man with the bad handwriting had gotten things wrong. At the very least it was the opportunity to visit the site where a stuffed Labrador Duck had resided before its fiery end.

Mainz is a community of 200,000 residents. The trip southwest from Frankfurt was on one of Germany's less zoomy train lines. Not nearly as stylish as an ICE or an EC, an IC, or even an IR, it wasn't quite as far down the list as an S-Bahn, but it was certainly in the second half of the alphabet. The train stopped at the Frankfurt airport and a lot of little communities where no one wanted to get on or off. The one fellow who detrained at Raunheim seemed frightened. Perhaps it is the site of a Cold War nuclear waste disposal facility.

Unlike the region around the main train station in Frankfurt, rich in drug addicts and sex shops, the region around the main train station in Mainz was rich in pastry shops and schoolchildren on day trips. Instead of using the small map of Mainz that I had ripped out of my guidebook, I chose to follow a pulse of children on the assumption that they were heading somewhere wonderful. This didn't work out at all, partly because the teacher was having trouble keeping the children all together, so that both the head and tail of the group were moving at a treacle pace, with the middlemost bits pulsing forward and back. I finally consulted my map, looked at a few street names, and then pulled out my compass. The task was compounded by roads with their own festive sense of direction, street names that changed every 50 yards, and a sudden and unexpected reversal of the Earth's magnetic poles. My map could not possibly have been right. The streets of Mainz may be navigable if you were born there and had never left. They may be just the ticket to confuse an invading army of foot soldiers, but they were completely useless to me on my first hour in town.

A kindly-looking gentleman getting ready to cycle away watched me with sympathy. Or possibly scorn. When asked, he graciously got back off his bicycle, pulled out his reading glasses, and peered at my map. After more than a minute of peering he was able to use my pencil to mark where I was.

After passing down some improbably minor streets, and rechecking my orientation at every intersection, I found a building labeled *Naturhistorisches Museum*. In fact, I found two. One looked likely to house the public display, and the other to house administrative offices. I entered the latter. I explained to a very helpful receptionist that I was looking for Dr. Ulrich Schmidt. She said that they had no one by that name, but that they did have someone named Dr. Ulrich Schmidt. My German accent must be absolutely awful. I asked if I might see Dr. Schmidt, but was told that this would be a problem because he was in Rwanda. Surely I misheard. Instead, she took me to see the Assistant Director of the museum, Dr. Herbert Lutz, a scholarly-looking fellow with speckled gray and brown hair and beard, who took my business card and read it from start to finish. My speech about Labrador Ducks and the reason for my visit to Mainz seemed to genuinely interest him, which was particularly gracious considering I had dropped in without warning (at least to him) and was asking for his time in what was likely a busy workday.

Lutz started to fill in the gaps in my knowledge. Dr. Schmidt was indeed in Rwanda. The museum had been working on ties with colleagues in Rwanda for quite some time. Personnel from Mainz had trained two Rwandans in curatorial skills, but both had been killed in the recent war. Despite horrific setbacks, the Rwandan museum of natural history was ready to throw open its doors to the public and Schmidt was there to help. Lutz went on to explain that the gentleman with the really bad handwriting who had written to Hahn was named Stadelmann, who served as a museum director of sorts in the years following the war. In the spring of 1945, like just about everything else in Germany, the museum had been hit by bombs, reportedly the night before evacuation of the natural history collection was due to begin. The bombs completed their intended job of making a hellish mess of the building, destroying much of the collection and its records. However, anything that could be salvaged from the wreckage was set aside. With a charred foot over here, a blackened fossil over there, and almost no surviving paperwork, sixty years later the museum was still trying to figure out exactly what was what. Lutz explained that by 1959 it was unlikely that Stadelmann had come to terms with everything that had survived the bombing. There was a chance that the Labrador Duck, or bits of it, might still be in the collection. Quite likely the museum had no written record of having ever owned a Labrador Duck. Lutz took a photocopy of my letter from

Stadelmann to Hahn and we went in search of the person who could give me a definitive answer.

Lutz and I left the administrative building, and using a back door, entered the building used to house the public display. Constructed in the fifteenth century as a monastery, after World War II it had been rebuilt rather cheaply and was now in need of a complete overhaul. This is all very well, but the people who hold the purse strings at Mainz City Hall didn't see the museum as a big priority. It was much lower in importance than the city's annual carnival, for instance. I could see where paint was jumping off the walls, and the ceiling had goopy water stains.

My head was spinning in anticipation. Was it possible that I was about to rediscover a precious specimen thought to have been destroyed sixty years before? Could it be that the stuffed Labrador Duck was just waiting for me to come along and spot it? A Mr. Hildebrand listened to the story in German. Lutz got his response in German. I practiced my look of casual aloofness and probably failed miserably. At this point, I really wish I could report to you that the Labrador Duck, Great Auk, and three Passenger Pigeons were sitting in the corner of a basement, a bit dusty and a bit smoky, but otherwise in fine shape. Regrettably, I can't. Hildebrand explained that these birds had been well and truly incinerated, with not a trace remaining.

I was provided with a ticket labeled *Freier Eintritt*, and so I got to enter yet another museum without paying. Unlike some of the world's great museums that I had visited, and with the deepest respect for the people working at the Natural History Museum, who were shackled by a too-small budget, this one was in pretty ratty shape. One can do only so much with a fifteenth-century monastery, but this wasn't it. The paint scheme was straight out of your grandparents' kitchen, circa 1955. Lightbulbs were burned out; radiators were exposed; some of the explanatory labels had been constructed decades before on a manual typewriter. I saw little artistic flair. Most of the displays left me asking why in the world I, or anyone else, should care. Room 12, labeled *Heimische Tiere*, was probably the best of the lot, with some effective, if dated, displays of animals arranged by habitat type, including field and stream, woodland, and creatures to be found near the house. Inexplicably, one end wall was covered with a hundred mounted heads of horned animals. Or just the skull and horns. Or just the horns.

This is not to say that the museum has nothing to be proud of. It has three stuffed quaggas that survived the war when so many other specimens were lost. The combination of stallion, mare, and foal represents a good chunk of the world's entire collection of this species of extinct horse-cum-zebra. I was told that there was now interest in genetic analysis to determine if the foal was the offspring of the adults on display. These three specimens are housed adequately, but whoever is in charge of finances for the rest of this museum should be ashamed and embarrassed.

Disappointed by the museum, but determined not to be disappointed by Mainz, I set off to discover something about it. I ate my take-away lunch in a small square in the shadow of St. Christoph Gutenberg Pfarrkirche. Dating to the ninth century, the church is dedicated to St. Christopher, but is better known as the parish church where Johann Gutenberg was probably baptized. You will probably remember Gutenberg as the whiz kid who invented the printing press. According to the people at the Association for the Beatification of Gutenberg, the Mainz hero began life as Johann Gensfleisch but changed his name to Gutenberg, feeling that "Beautiful Mountain" was a more dignified surname than "Goose Flesh."

I found the tourist information office, despite a series of signs that pointed me to the middle of the Rhine, the city's crematorium, and the transit of Venus across the Sun. The woman behind the desk provided me with an annotated map of Mainz. Out of English versions, she was able to provide me one with marginal notes in French. She circled the *Dom* and the church of St. Stephen as places that absolutely should not, could not, must not be missed. I baited her. "I understand that there is a natural history museum in town." "Yes," she said, "It's . . . it's right . . ." She couldn't immediately find it on the map. "Is it any good? Is it worth a trip?" Diplomatically, she neither lied nor warned me off. "Yeah . . . well . . . well . . . We all had to go there as children, but . . . yeah . . . it's okay." Mainz city councillors should hear this sort of halfhearted endorsement.

I strolled to the river. What a river it is. Romans had been sensible enough to build a settlement where the Main meets the Rhine.

The Romans knew a river when they saw one; the Rhine runs more than 800 miles from the Swiss Alps through Liechtenstein, Austria, France, Germany, and Holland before emptying into the North Sea. I was delighted to see that, unlike so many other great rivers in Europe, this one had some actual commercial traffic. Each barge that chugged by had a single gray automobile on deck. A little less useful than a lifeboat, I would have thought.

Marching toward the center of town, I found the *Dom*, although it looked a lot more like a fortress than a cathedral. The windows that pierced the outer walls are the sort seen often in British castles, designed to make it easy to fire arrows down on potential invaders. About two dozen people had seated themselves in pews on the right side of the nave, but no one was seated on the left side. Risking some horrible faux pas, I plunked myself down on the left side. From my vantage point, the *Dom* seemed to be saying: "This monument to our praise of the Father, the Son, and the Holy Spirit will still be standing long after the memory of humankind has passed from the Earth." Parishioners may be bathed in warm light on a bright Sunday, but it was all a bit bleak the day I visited. Much of the construction had been completed in dark red sandstone, with dark marble floors and an acute shortage of windows. There was also a surprising paucity of tributes to God.

At the entrance to the *Dom*, there was a large sign written in German, English, Spanish, Italian, Sanskrit, Croatian, Djiboutian, and !Kung click language, explaining that, as a house of worship, a degree of decorum in style of dress was appropriate. As I sat contemplating the nature of my immortal soul, and why the Dom's columns were square instead of round, a couple off to my left got into an argument. They were dressed in matching cut-off jeans, T-shirts, and denim baseball caps. The argument, loud and prolonged, concerned a message displayed on their digital camera. "Look," she shouted, "it says right there, 'Picture is blurred.' " "Well, I don't care what it says," he replied. "For what we paid for that camera, it had better be in focus!"

The Gutenberg Museum, Mainz's most popular museum, was just down the way. The people of Mainz are very proud of Gutenberg, and very proud of God, presumably not in that order. After all, there is only one Gutenberg Museum in Mainz and lots of really big churches. Without Gutenberg there would be no cheap erotic novels, but without God, reasoned the citizens of Mainz, there wouldn't be much of anything. Not having gone to the Gutenberg Museum in Strasbourg the previous summer, in a sense of fair play, I didn't go to the Gutenberg Museum in Mainz.

EVERYONE IS ENTITLED to a bad day now and then. Once in a while, for reasons completely beyond your control, a day goes wrong from beginning to end. On a morning with an important meeting, your water heater dies without warning and your hair must be washed in cold water in the kitchen sink. Your cat gets sick on the carpet. Your nose begins to bleed for no good reason. Pixies move your car keys from the hall closet where they are placed carefully every day to a dark corner of the laundry room. Not that keys would do you any good because your car has a flat, and the spare tire, checked just last week, is now inexplicably flat too. There is no predicting such days; these things just happen, and you comfort yourself by humming tunes from Broadway musicals. You are a nice person and, being a nice person, you don't want to make anyone else miserable. But try as you might to keep it all to yourself, your bad day takes on a life of its own, and when you finally get to work, you pass it along. And as long as it doesn't happen too often, no one really minds.

But when bad days happen to the same person again and again, and that person makes an art form out of passing along their bad mood, a reputation is in the making. Some miserable people lose their friends, and some lose their jobs, but some manage to hang on and pollute their immediate world. In some cases the poisoned atmosphere can linger, even after the poisoner has moved on to new challenges. This is my theory to explain why some workplaces have a reputation for being uncooperative and inefficient year after year. Everyone knows it, but no one knows how to stop it.

The Zoologisches Institut der Universität Tübingen was digging itself out from under a reputation for being uncooperative and inefficient. When I had contacted the museum about ten years earlier, asking about their Labrador Duck, cooperation and efficiency were nowhere to be seen. The response to my first letter warned me that the curator was extremely reluctant to respond to inquiries because of "internationally organized gangs" that had stolen specimens from the museum. I passed along contact details for well-placed persons in the world of ornithology who could vouch for my honorable character, but this effort did not elicit any response. The response to another request two months later explained that I could expect help shortly. In this case, "shortly" meant "never," and, in the end, I couldn't even get an acknowledgment that the museum really had a Labrador Duck. Luckily, all of that had changed in the interval, and my new contact, Dr. Erich Weber, was being entirely enthusiastic and cooperative.

The trip from Frankfurt to Stuttgart was on a state-of-the-art ICE train. The interior was painted in soothing shades of blue, green, and aubergine. The WC had a lovely floral scent and was nicer than any I have visited on an airplane. At the end of the carriage, an LCD display showed the anticipated time of arrival and our current speed. My handwritten notes read: "The top speed I saw was ~~161 km/hr!~~, ~~183 km/hr!~~, ~~196 km/hr!~~, ~~212 km/hr!~~, ~~234 km/hr!~~, 250 km/hr!" Some of my fellow passengers munched on pastries or pretzels, but two gentlemen across the aisle toasted each other's health with goose liver pâté on crackers and a big bottle of champagne. If airline travel were anything like this, I would do a lot more traveling.

The train from Stuttgart to Tübingen was less sophisticated, but no less fun. I traveled with seven students from the Atlanta university system, plus one fellow who wanted to make it absolutely clear that he was from Columbus, Ohio, not Atlanta. They were a delight to be with. Enthusiastic, polite, engaged, and forward looking. Bound for Heidelberg, but with a long layover, they had decided to fill the time with a side trip to the university town of Tübingen. They were particularly keen on Tübingen because of its ties to Goethe, who drank heavily and vomited wildly there in the late eighteenth century. They politely asked about Canada, and showed at least a passing interest in my duck quest. I was polite too, pretending that I knew all about Goethe.

I really, really wish I could have spent the night in Tübingen. The city of 87,000 residents is usually described in terms of the grace of medieval stained glass, cobbled alleys, and half-timbered houses, all enriched by the presence of its enthusiastic university student body, which swells the population by 22,000. No inconspicuous regional college this, it employs every eighth person in Tübingen. Count Württemberg established the university in 1477, after a pilgrimage to Jerusalem. His vision was of an institution to help "the world to drink comforting and healing wisdom and thereby extinguish the pernicious fire of human ignorance and blindness." These seem like particularly lofty goals considering that, for many students, their time at university is mainly an opportunity to move away from home. An overnight stay would have given me a chance to see some of this, and more time to work on my pronunciation; Toob-in-ghen, not Tube-in-jen.

My first goal at Tübingen's *Hauptbahnhof* was to unload my backpack. I couldn't find a checked luggage desk, so I scouted for lockers. None was really big enough, but I crammed my bag into a locker that asked me for €1.50. The locker told me that it would accept coins in any combination of €0.50, €1.00, or €2.00, but that I wasn't to expect any change. Fair enough, but no matter what combination of coins I gave it, my offering wasn't satisfactory. I tried another locker, with the same result. Figuring that I had to be doing something wrong, I asked a train station employee for help. After hearing me speak German with an accent, he shrugged his shoulders and walked away. In desperation I tried creative combinations of British pound coins, Canadian quarters, and American nickels, but my locker loved me none the more. I thought about stuffing my chewing gum into the coin slot, but didn't. It looked as though I was going to be stuck with my backpack.

Dr. Weber had sent me an email describing my options for getting to the university. If I didn't want to walk, the number 5 bus would take me straight to campus. I found the number 5 bus pulling away from the train station, and leapt on, my backpack, briefcase, and duck-measuring kit trailing behind me. I asked the driver to confirm that the number 5 did, indeed, go to the university, and he replied, "*Nein.*" The passenger ahead of me turned around as though keen to help, and I asked her if the number 5 would take me to the university, and she said, "Ja." I split the difference, and decided that the bus would get me somewhere close to where I needed to be. The "Ja" lady then asked if I could change a five-euro note, as neither she nor the driver had any coins, and tickets were to be purchased from a coin-operated machine on board. I didn't have enough change for both of us. A lady further back in the bus was able to provide change, and showed the "Ja" lady how to purchase a ticket. "Would you mind showing me, too?" I asked.

By this point, I was dripping with sweat from the exertion of carrying all of my luggage and the fear that this bus was going to deliver me many miles from my destination. I broadcast a question, in English, to my fellow passengers. "Would someone please tell me when we get to the university?" One very kindly young woman, taking pity on me, looked at my little map and said that she would get off at my stop and escort me where I needed to go.

By the time I got to Weber's office, I was spewing sweat at an Olympic rate. This seemed to concern Weber, particularly since he thought the day was on the cool side. I laughed it off, explaining that we Canadians thought anything above the freezing point of helium was too warm. He looked at me as though I might be in the throes of advanced malaria or perhaps leprosy. Or insanity.

Despite my potential for contagion, he took me to the basement, to a preparation room with fume hoods and cluttered workbenches, and set me up at a trolley. The bird collection is an interesting one. Founded around 1841, it is a part of the Zoologisches Institut der Eberhard-Karls-Universität Tübingen, and includes something like 2,000 skins and mounts, 1,300 partial skeletons, and 500 sets of eggs. These were all locked safely away, but the Labrador Duck was waiting for me.

Labrador Duck 20

As I prepared to get down to my poking and prodding, Weber explained that my duck came from the collection of Herzog Paul von Württemberg, who was responsible for much of the early bird collection. He had been a wealthy hunter who had traveled extensively in search of things to kill. This isn't to say he necessarily shot the Labrador Duck himself; a member of his party might have nabbed it, or he may have purchased it on his travels. And just because the tag around my duck's leg says that it had been collected at *Hudsonbai*, this shouldn't be taken as unimpeachable evidence that my duck had necessarily ever been anywhere near Hudson's Bay.

Weber also explained that this specimen, remounted in the 1950s, had probably been a study skin before being remounted as a taxidermic preparation. Therefore, this drake, an adult, probably had brown glass eyes only because the preparator had a bunch lying around at the time. He asked me if I needed any feathers for DNA analysis. "Why? Do you have some to spare?" A few small feathers had come out in cleaning, and these were in a ziplock bag attached to the base. The vandal in me wanted to take one or two as souvenirs, but I resisted temptation, as I knew that a thorough DNA analysis of Labrador Ducks had already been completed.

The museum's current preparator, Jürgen Rösinger, asked me how this specimen ranked as Labrador Ducks go. Well, let's see . . . He has been repaired in spots, but those spots are fairly inconspicuous. He was shedding some of his filling through a hole in his flank. He wasn't particularly dirty, and no one had felt inspired to apply too much paint to his feet or beak. His base is covered in sand, which is unique without being gaudy, and his posture is a little out of the ordinary. His right wing hangs a bit lower than it did in life. All in all, he is somewhere near the middle of the pack. The drake, that is, not Rösinger. Rösinger was quite near the top of the pack.

Having finished my peek and poke, I asked Weber to recommend a place for lunch, and he took me across the way to a student cafeteria. The hall was nearly full, as students were still immersed in their classes before being given a break until mid-October. I tried to pay for lunch, but Weber insisted on catching the bill, explaining that I was his guest. The last of the museum's bad reputation died on the spot. We chatted for a spell about common interests, including the taxonomy of birds and the pressures of a university position based largely on teaching. Given his teaching load, Weber has precious little time for research. It seemed to me that he would benefit from a couple of months in the sunshine chasing birds. We compared our two universities. His was founded in the fourteenth century and had 22,000 students. Mine was eight years old and had 600 students. At my university, students paid about $4,000 each year in tuition. At his, students paid a small registration fee and no tuition at all. There had apparently been a student revolt when the registration fee had been increased to €80.

Probably fearing a return of my unique commitment to sweating, Weber offered me a ride back to the train station an hour hence, suggesting that I might want to tour the department's small public museum and could chat with his long-time museum associates, Herr and Frau Mickoleit. In contrast to Rösinger's punky black haircut and punky black style, the Mickoleits were severe. Both sported vigorous haircuts, and both spoke in a clipped, intense manner. They told me a lot about the history of the collection and how it came to be where it is now, but a lot of it whizzed by before I could take any notes. Herr Mickoleit implied that he had been at the university for a great many years, as man and boy.

They took me to a locked room where particularly valuable specimens were displayed. Students were allowed to visit, but it is normally locked up because of fears of theft, presumably by internationally organized gangs. Herr Mickoleit was particularly proud of a Passenger Pigeon, a Carolina Parakeet, two Ivory-billed Woodpeckers, and a thylacine. At the far end of the room was the most eye-catching display of all. It was a longitudinal section through the head of an elephant, a couple of inches thick, housed in a huge glass case filled with foul-looking preservative. To obtain this bizarre apparition, an elephant's head had been frozen in a block of ice, taken to a quarry, and sawn through with a diamond blade.

With a few minutes to spare before my ride to the train station, I took in the zoology museum used by the university's students. Like the museum in Mainz, it was plain and had been constructed on a small budget. Unlike the museum in Mainz, in some intangible way the display of birds, mammals, and insects made me care. There was a bit of flair to each display. Perhaps this should be the first stop on the grand tour of small museums by the bean counters in Mainz.

AT THE PLOCHINGEN train station, the itinerary generated by Deutsche Bahn gave me only six minutes to get from platform 59 to platform 4. Luckily, the Plochingen train station isn't a very big place, and the platforms were only 500 feet apart. Leaping on the Munich-bound train with two minutes to spare, I found it completely packed.

It is a good thing I had reserved a seat, or I might have been seated on the roof of the smoking car—facing backward.

A lady named Eva sat in the seat beside me, and proved to be an absolute joy as a companion. She, her physician husband, and their three young children were on their way to a wedding at a monastery on a lake east of Munich. Like many Germans, she spoke fluent English but seemed flustered every time she hit some rarely used expression that she couldn't translate into English. She put her rusty English down to a period that she had spent in Strasbourg, polishing her French. I feel such an idiot when people tell me things like that. I am fairly fluent in the language of birds, but that isn't much help when a message comes across the public address system of the train, as it just had. Eva told me that the message explained why we were still sitting in Plochingen when we should have been fifteen minutes on our way to Munich. While the train was pulling into Plochingen, a woman had either suffered a seizure or gone into a coma (Eva's English failed her here). Her husband had assisted until help arrived at the train station.

Even Eva's children were getting in on the linguistic act. Attending some sort of international school, they were rapidly becoming quadrilingual. Hearing that Eva's youngest daughter was celebrating her fourth birthday, I pointed to her, smiled, and said, "*Eins, zwei, drei . . .*" "Four," she replied, looking at me as though I were a dimwitted circus clown.

THE ZOOLOGICAL COLLECTION in Munich was founded in 1759 as a private collection of Kurfürst Maximilian III Joseph von Bayern. The collection really took off only after Johann Baptist Ritter von Spix arrived in 1807, and then, after a six-decade slowdown in bird activity, picked up again with the arrival of C. E. Hellmayr in 1903. The collection now belongs to the Bavarian government, and Zoologische Staatssammlung München translates as the Zoological Collection of the State of Bavaria. One of the highlights of the museum is its collection of 6 million butterflies, which probably rank it number one in the world. No prizes for correctly predicting that they also have a Labrador Duck.

The walk from the hotel to the museum was less than a mile, but by the time I drew close, I had resumed my routine of freestyle sweating. As I passed one of Munich's 800,000 *Apotheke* shops, I discovered why. Large weather dials showed that the temperature had made a healthy commitment to the mid-90s, and the relative humidity was slightly higher than I like it in the shower.

The museum building that houses Munich's natural history research collection is a curious structure, with much more belowground than above it. Located in the suburban outskirts of northwestern Munich, if you didn't know what you were looking for, you would be excused for thinking that the place was just an abandoned construction site. The Labrador Duck and many of his contemporaries survived World War II by being moved out of the city center to buildings in outlying regions. The old museum building and all specimens that had not been evacuated had been destroyed. At the end of hostilities, the collection had been moved to the Schloss Nymphenburg. But more than twenty-five years ago, the director general of the facility suggested that a new facility be constructed to house the research collection, while the public displays remain in the castle. As I approached the subterranean facility, an air vent brought me the faintest whiff of mothballs and other preservatives, but I am sure very few residents of the neighboring community even noticed.

Labrador Duck 21

A receptionist buzzed Josef Reichholf, an ornithologist and head of the museum's Vertebrates Department. Reichholf sent his assistant to fetch me. As she escorted me to the duck, she said, "My English is not good." "*Mein Deutsch ist schrecklich,*" I replied, and felt rather clever for getting that much out. The duck, a taxidermic preparation without a base, was lying on its side in a hermetically sealed plastic bag, and Reichholf's assistant found a pair of scissors to free it. I immediately knew more about the duck than Paul Hahn ever had; he knew only that it was a male, but I could also see that it was an adult. Reichholf appeared, and we made pleasantries about my trip to Munich and my early impressions of the city. Noting that I was sweating like a pig at a luau, the assistant opened a big window. If we had been in one of the many rooms situated belowground, I probably would have quietly expired.

I got down to work. A few minutes later, the assistant came back to offer me coffee. A few minutes later, Reichholf came back to offer me coffee. A few minutes later, a custodian arrived. She proceeded to empty the wastepaper baskets but didn't offer me coffee. She then got down to a serious assault on the work surfaces around me. She squirted liquid soap and turned the water faucet on a little higher than I thought prudent, given the proximity of the valuable duck.

The specimen is a good one, although his head is a bit loose. It was the first specimen I had seen that had been given cherry-red glass eyes. In the long term, he would probably be better off if he were attached to a new base. Lying on his side, his feathers were getting a bit pressed out of shape in places, particularly along his crown.

When I finished up, without stealing any of the loose feathers in the bag, I found Reichholf in his office, and asked if he could clarify a few points. He told me that the museum had about 60,000 specimens of 6,000 bird species. I asked about the initials H. v. L. on the duck's tags. He dashed off to find a book. When he returned, Reichholf was able to tell me that the initials stood for Herzog von Leuchtenberg, who had originally been called the Duke du Beauharnais. A stepson of Napoléon Bonaparte, he married a Bavarian princess. Since no one in his adopted land could pronounce Duke du Beauharnais, he changed his name. The Labrador Duck had been in the collection of von Leuchtenberg when he lived in Eichstätt, a small community between Munich and Nürnberg. From there the duck entered the collection of the state of Bavaria. Considering that Hahn's complete description of the Munich duck was "male," I think that I had done pretty well in gathering new information.

I then asked Reichholf about the museum's holdings of Great Auks. At one time they had apparently owned two. When Errol Fuller wrote his magnum opus on Great Auks, he had encountered considerable difficulty getting much information out of this museum. Reichholf was able to confirm that the museum had only one specimen, and that the second specimen of song and legend was nowhere to be found. Although the remaining specimen wasn't in the best shape, they were keeping a close eye on it to make sure it didn't get any worse. Reichholf had a meeting to dash off to, but his assistant took me to room M110, labeled *Magazin Ornithologie,* to see the Greak Auk. As I prepared to snap a couple of photographs, the assistant turned the bird so that its best side showed.

I wandered back to my hotel room for my second shower of the day. As I dried off, I considered my options for my remaining time in Munich. I could rest and prepare myself for the latter half of my German expedition, but that wouldn't give me much to talk about when I got home. Alternatively, I could take the train into the heart of Munich, visit its official tourist sites, get really, really hot, and wear myself out. Given how tired I was from the day before, I favored a third option, bound to be a lot less strenuous, but still a bit of fun. My cab driver had told me the Schlosspark Nymphenburg was an incredibly beautiful park. It wasn't too far from my hotel and it would give me the chance to see where the Labrador Duck had lived from the end of the war until being moved to its current home twenty years ago. Who would blame me for taking the third option? Surely Munich is just another European city, with a central core pretty much like any other.

But as I got closer and closer to the park, I also got closer and closer to the Obermenzig *Bahnhof*, and the train that would take me to the city center. An odd sort of guilt set in. The problem was that Munich was not just any other European city. People lived and worked and loved and procreated and died there, and were probably very proud of what they helped to create anew each day. And so, despite the smell of vomit at the train station entrance, I took the stairs to the platform.

Most visitors to central Munich walk east from the *Hauptbahnhof* along Bayerstrasse, but I found that by walking along Schützenstrasse instead I could see some much better architecture and miss Europe's most extraordinary gambling facility, several beggars, and a Frankfurt-style live sex show theater. With this one exception, the region east of the Munich train station could not have been more different from the region east of the Frankfurt train station. Instead of prostitutes and drug addicts, I found a fountain that was so fabulously attractive that every stone seat around it was occupied by a least one bottom, with a small flock of people hovering, waiting for their turn to sit. I passed through a huge stone arch and found an array of shops that included a McDonald's, a Foot Locker, a Benetton's, a Gap, a Vodafone outlet, and a Swatch store. Without being too jaded about retail opportunities, I did find some really odd ones, including a Salamander shoe store, Wormland men's clothing shop, and Christ jewelry. I have visited a lot of great art museums, but I don't recall seeing a painted image of the Son of God wearing a lot of bling.

I found my way to the very heart of the Old Town, the Marienplatz, with its central Mariensäule, a column topped by a golden Virgin Mary. Mary was carved in 1590 but didn't find her way to the top of the column until 1638. It had something to do with Germany beating Sweden in a war. Toward the base of the column, protecting Mary from nasty things that might want to climb up, are four helmeted, winged, cherub warriors with spears and swords. The first is slaying a dragon, the second a lion, the third a serpent, and the fourth a chicken. Perhaps Mary was allergic to feathers.

From there I wandered over to St. Peterskirche and climbed to the viewing platform at the top of the Alte Peter. A lovely breeze helped with my latest bout of sweating, and also helped to drown out the wheezing of the less active component of the visitors who had climbed the 297 steps for the view. I counted seven clock towers, including the one I was in, and was charmed when they struck the hour in near-perfect unison.

The S-Bahn 2 line took me back north to the region near the museum and my hotel. Although in need of my third shower and a good lie-down, I heard a nagging little voice. It was telling me I had no right to leave Munich before I saw the Schloss Nymphenburg. The castle had begun life in comparative modesty as the villa of the Electress Adelaide of Savoy, but over the next century became grander and grander, and served as the summerhouse of the royal family. But, more importantly, von Leuchtenberg's Labrador Duck had once lived there. And so, instead of turning right at the train station, I turned left and aimed for the park.

Entering the park from the north side, I wandered along a path through a lovely deciduous forest. A good spot to cycle, and an even better place to run, it was probably a great place to canoodle with people you shouldn't be canoodling with. Just when I had had about enough of the forest, I came out into a magnificent garden in the style of an eighteenth-century English estate, with wide lawns, a fountain in full spray, and a pond full of giant carp and covered with coots, ducks, geese, and swans, bordered by statues of all the coolest Greek gods. All of this was just the view from the east side of the palace. Once I passed through a portal to the west side, the scope of the whole affair was revealed. There was another majestic fountain and more waterfowl, swimming in grand canals. The castle had two additional wings that were so far apart, each had its own clock tower.

The north wing held the Museum Mensch und Natur, the public display portion of the Staatliche Naturwissenschaftliche Sammlungen Bayerns. I was on the verge of admitting that I had seen all of the natural history museums I could handle, but parted with €2.50 for another experience. This should be stop number two on the tour of museums by the tightwads in Mainz. The displays were interactive at every turn. There was an abundance of spinning models and flashing lights, with handles to turn and buttons to push. On that Friday afternoon, the children on hand were lapping it all up. If the folks in Mainz modeled their museum after this one, children would see the visit as the highlight of the school year. Even though there wasn't a single word of English, I usually got the point. I was particularly touched by a display on thylacines. Beside the stuffed beast was a television screen showing black-and-white movie footage of the last-ever thylacine, a captive, pacing back and forth in its cage. It looked so horribly lonely, and so horribly sad, that I turned away in shame at its extinction.

It was time for my next city and my next duck.

WHEN THE THERMOMETER is clinging tenaciously to the mid-90s, I can see the appeal of eating outdoors, as could many Berliners on an early Saturday evening. The sidewalks around my hotel were crammed with dining opportunities, and after scanning about twenty posted menus, I settled on a place that served me an eggplant and mushroom dish in coconut milk sauce. Several passers-by called out, "*Bon appétit!*" to which I responded, "*Merci beaucoup!*"

I knew Berlin's (take a deep breath!) Museum für Naturkunde, Zentralinstitut der Humboldt-Universität, Institut für Systematische

Zoologie to be home to one of the most beautiful Labrador Duck specimens, but while dining, my mind was not fully occupied by thoughts of ducks. I couldn't help but reflect on what a truly astonishing city Berlin is. In Europe, it seems that even the smallest communities can trace their histories back to the cooling of the Earth's crust. In contrast, Berlin wasn't even established until the thirteenth century, and then only as a small fishing village. It plodded along for two centuries without much fanfare until the Holy Roman Emperor instructed the House of Hohenzollern to give it a boost. Italian artists moved in and coffee sales soared. Rats moved in and bubonic plague struck three times in twenty-six years. As if that were not enough, the Thirty Years' War left only 6,000 survivors. Keen to get its numbers back up, Berlin invited foreigners to settle, including wealthy Jewish immigrants from Austria and Huguenots fleeing France, tripling the city's population in just forty years. The city became the capital of Brandenburg-Prussia in the eighteenth century, amid general cheering and celebration. Canals were dug, castles built, fortified walls erected, and lime trees planted. After all the fanfare, the city had the living daylights kicked out of it by Napoléon a century later.

After the Emperor's defeat at Waterloo, Berlin got revved up all over again, with the construction of railways and the establishment of Humboldt Universität. Then there were hunger riots. And a war with Denmark. And a war with Austria. And a war with France. When the German Empire was proclaimed in 1871, Berlin was installed as its capital. Being named capital of a country with expansionist goals seems to me to be similar to having a target painted on your back. World War I was not a party, resulting in famine and strikes, but that wasn't a patch on World War II. Its citizens faced food shortages, British air raids, and general death and devastation. The war just wouldn't go Berlin's way. At the tail end of World War II, 1.5 million Russian soldiers came to town, and they weren't in a festive mood. Berlin became the bull's-eye during the ensuing Cold War that simmered between the United States and the USSR. The Berlin Blockade of 1948 didn't help. Nor did the Berlin Wall, constructed in 1961. Reunification of East and West Berlin wasn't cheap, leaving the city today with a debt in the tens of millions of euros. Berlin just seems to be the sort of place that should have folded the tent flaps a very long time ago.

And yet, despite all of this, Berlin is home to 3.3 million people—greater than the combined populations of Mauritius, Fiji, Luxembourg, and Cyprus. The city has twenty-eight palaces and more than eighty museums. It is one of Europe's great cultural centers. In a feature article, *Time* magazine proclaimed that Europe needs a strong Germany, and Germany needs a strong Berlin.

All of this left me feeling afloat. I had to wait one full day before I could examine the Labrador Duck, and in that time I wanted to be able to come up with something insightful to say about Berlin. There must be a small army of scholars who spend their whole professional lives considering the history and future of Berlin, and even they must mess it up once in a while. Surely it would be folly to think that I could encapsulate even the smallest corner of Berlin in the time at my disposal.

And so, over my second lager and the last bits of dinner, I set myself a more modest task. Most great cities have one or two enduring images, good or bad, that conjure up that community in the minds of people everywhere. Think of the Eiffel Tower, the Coliseum, the British Houses of Parliament, and the fault line under San Francisco. Enduring images do not have to be specific buildings or structures, as evidenced by Mardi Gras and the Latin Quarter, or the Tiananmen Square protests. With a free Sunday, I set myself the task of finding something that would serve as my enduring image of Berlin. When I think about Berlin, fifty years hence, I want a single image to come to me as I sit in my rocking chair at the Old Boys' Home.

I STARTED VERY early on Sunday morning, and walked south from my hotel room to the Fernsehturm, a gigantic telecommunications tower. At precisely 365 meters, it is the second-tallest freestanding structure in Europe. It has an observation platform at 203 meters, and a revolving café for those who feel that coffee tastes better when spinning. Some locals apparently call the structure *Telespargel*, a stalk of asparagus. If a few thick cables were added near the top, it would look almost exactly like a giant bacteria-consuming virus, impaled one-third of the way along a giant spike. It was too early in the morning to be admitted to the tower, and so it failed as my enduring image of Berlin.

If the Fernsehturm were to fall over, it would probably hit the Marienkirche. The house of worship is full of really neat things like a carved altar from 1510, a baptismal font supported by black dragons, an alabaster pulpit, and a Gothic wall fresco called *Totentanz, The Dance of Death.* Regrettably, I didn't get to see any of these things. Being a Sunday morning with scheduled services, it would be a few hours before it would be open to visits by non-worshipers. From the outside, the church was pretty enough, I suppose. Perhaps a bit austere for my taste. For me, the best thing about the church was the flock of tits all over the brickwork, each looking for a meal.

The Neptunbrunnen is a beautiful fountain inspired by fountains in Rome and Versailles. Neptune is held aloft on an open clam shell by hybrid horse-men with duck feet instead of hands. Neptune is held above a serpent, a crocodile, a seal, and an octopus with seven legs. A septopus, I suppose. The fountain is probably even better when it is turned on. Dry, I didn't think it could serve as my enduring image.

Crossing to Museum Island, I was drawn to the Berliner *Dom*. Whatever points the church loses for architectural simplicity, it makes up for in tenacity, having been rebuilt after severe bomb damage in the war. As parishioners filed in, I sat on a park bench across from the main entrance, listening to beautiful voices singing God's praises, as they wafted over the roar of four idling tour buses parked just down the street. Most of the men arriving to join in prayer were dressed in black or gray, but many of the women had chosen bright, cheery outfits, which seemed suitable for celebrating God's gift of life on a bright, cheery Sunday morning. At 9:37, the *Dom*'s bells began to peal. Something told me that this wasn't a typical Sunday-morning Mass. Perhaps it was the flock of police officers. Perhaps it was the television crew setting up on the sidewalk. The cheesy part of me hoped that I was going to see a celebrity wedding. At 9:47 the bells ceased and I turned away.

The Lustgarten awaited me. The site had been used to grow vegetables and herbs until taken over in the seventeenth century by the Great Elector, who apparently favored pleasure over nutrition. I walked through the grounds, keeping myself open for anything that would instill lusty thoughts. Nothing did. I continued past ever so many museums on the island, but I didn't have the impression that I would find my enduring image inside any of them. In my ideal world, where admission to all museums is free, Museum Island will be in line for a big change, and I suppose that isn't likely to happen until Berlin crawls out from under its crippling debt.

I came to the Gertraudenbrücke, a bridge adorned by a bronze statue of St. Gertrude, the patron saint of a hospital long since demolished. The statue, rich with symbolism, has Gertrude leaning over a poor boy, offering him a lily (symbolic of virginity; his or hers, I don't know), a flagon of wine (indicating love), and a distaff (for charity). The mice at the base were probably symbolic of rodents in general. The boy was using his knee to pin a rather distraught-looking goose. Next to the bridge was the Galgenhaus (Gallows House), so named because, history tells us, an innocent girl was hanged there. There was no plaque describing the hanging, but signs indicated that parts of the building were for rent.

Carrying on with my search for simple charms, I discovered a café that provided me with a *bis' Weissbräu* and the day's special—*Gebackener Camenbert mit Preiselbeersauce*. I knew the word *mit*, and guessed that *Camenbert* was Camembert cheese, but I relied on my phrase book to bail me out on the other words. *Gebäck* are pastries, and *Preiselbeeren* are cranberries. Thoroughly vegetarian, it came with French bread and a salad based on clover. It was almost good enough to become my enduring image, but not quite.

When a planned meeting with the brother of a work colleague did not materialize, I went off in search of dinnertime adventures on my own. Wandering west along the Torstrasse, I found a series of Middle Eastern cafés and take-aways, and opted for a falafel sandwich to go. I felt a bit like part of the scenery, among the early-evening crowd, even though almost everyone was more appropriately dressed for the warm summer weather than I. And then, while walking along the Invalidenstrasse, a patch on my left thigh seemed strangely cool. Despite my best efforts to eat carefully, my last bite had caused a stream of tahini sauce to pour down my leg and onto my shoe. So much for being an inconspicuous part of the backdrop.

A few steps farther along, I came across a strip of rubble and grass with a bit of new construction on either side of the street. It seemed odd to have so large a chunk of valuable real estate so poorly used in a city of this size. And how odd that the site should be so linear. Could it be that I had found the site of a chunk of the notorious Berlin Wall? Little of the wall remains, and most of the material that had made up the wall has now been recycled in road construction. One thousand years from now, our ancestors will curse us for the wall's near-complete demolition. I understand the need to put unpleasant things behind us, but if they get put too far behind, we tend to forget them and run the risk of repeating them.

Heading back toward the hotel, I strolled through the Volkspark am Weinbergsweg. The park was packed on that early Sunday evening, and my stroll was leisurely, if only because there were acres of skin exposed to the fading sunlight, and some of it was worth looking at. In an era of depleted ozone and soaring skin cancer rates, some people had seen altogether too much sun that afternoon. One young lady had rolled up her white tube top so that just her nipples were covered. Her boyfriend caught my eye and scowled. I suspect he had been doing a lot of scowling that afternoon. A few dogs looked very warm. A few books looked very well read. A few fellows in Jamaican national colors looked very Rastafarian. A gentleman tending a cheap metal barbecue spouting big blue flames looked very perplexed. At lunchtime, I had been the youngest person in the restaurant. This evening I was the oldest person in the park, and the only person whose trousers were stained with tahini sauce. Could this be my enduring vision of Germany? A scene of young people enjoying their youth in a city park, as they did in so many urban centers? I could see nothing that made it an image of Berlin as opposed to an image of, for instance, Cincinnati.

I WAS OFF for the zoology museum early the next morning. It was a Monday, and the museum was closed, as most German museums are on that day. I was, however, a bit distressed to find that the main entrance was barred by a big metal gate, and there was no one in sight. I tried one side of the building and then the other without success, and then found a man in a traffic-control kiosk. "*Guten Morgen. Mein*

Name ist Professor Glen Chilton." I threw in "Professor" for a bit of oomph. "*Ich möchte gern Frank Steinheimer sehen.*" The traffic control guy looked at the lady he had been chatting with, and they both shrugged. I tried again in English, and they shrugged again. I ventured the word *Ornithologie,* and finally *Zoologie,* and was directed to the correct door. The museum's Keeper of the Keys offered to guard my bags and hailed Steinheimer.

This was the third time I had met up with Steinheimer. The first was when he had helped me on my initial visit to Tring. The second was the bird collection conference in Leiden. Each time I saw him he looked younger, more casual, happier, and more full of youthful vigor. If he doesn't stop this, he will soon be an irresistible woman-magnet. Steinheimer was dressed particularly casually that day because he had come in during his vacation time to help me and a pair of young researchers from the Czech Republic. He took me to the Labrador Duck, which, he explained, hadn't been brought out for examination in a very long time. Steinheimer then did exactly what I hoped that he would do—he offered the three of us a tour of the facility.

The Museum für Naturkunde is one of the largest natural history collections in the world, with more than 60 million specimens of mammals, insects, and plants, and representatives of almost all the bird species known to science. It was founded in 1810 as an expansion of the collection of Kaiser Wilhelm II. The list of contributors to the bird collection reads like a who's-who of early ornithology. In a refrain all too familiar, much of the collection would likely have been lost to wartime bombing but for the efforts of curators who hid specimens away in cellars, bank vaults, and village schools.

Clearly Steinheimer loves the Berlin museum very much. We started our tour in one room of the ornithology branch of the museum's scholarly library. Books were shelved floor to ceiling, and the ceiling was very high indeed. Ladders provided access to items out of reach of the spiral staircase and balcony. Given the value of the collection, I was surprised that we were allowed to amble in, and even more surprised that Steinheimer was allowed to handle the books without white cotton gloves. I certainly wasn't going to leave my fingerprints on them. He showed us examples of the first books ever devoted entirely to birds, dating from the fifteenth and sixteenth centuries, originals of books I had only heard of and had not seen even as reprints. The hand-colored artwork in some was both subtle and emphatic. Two of the best books were in need of rebinding, which would cost about €1,000. This seemed a small price to pay, considering that together they were worth about €300,000. Steinheimer claimed it would not take long to make a short stack of books whose combined value would exceed 1 million euros.

Steinheimer then took us through the collection of stuffed birds, which contained many items whose scientific value was inestimable. These had escaped demolition in Allied bombing by the narrowest of margins. We were shown a specimen, no longer more than a foot and a small bundle of feathers, with a shard of glass impaled in its wooden base. The glass had been part of a window before the bombs started landing. He showed us wooden cabinets with shrapnel still embedded. Some specimens were quite dirty, but considering the hell they had been through during the war and since, it is surprising that they remain at all. The one thing that we weren't able to see was arguably the museum's greatest treasure. There are only a handful of *Archeopteryx* fossils in the world, and Berlin has one of the best. This creature is usually seen as an intermediate between bird-like dinosaurs and full-fledged birds. I am sure that everyone has seen photographs of *Archeopteryx* fossils, and many museums own re-creations, but to see a real one would be near the pinnacle of my scientific experiences. The museum had recently cobbled together the funding necessary to put their fossil on display, but at the time of my visit, it was locked away in a safe, and unfortunately I didn't have a good enough reason to ask them to take it out.

The zoology collection is impressive, but the buildings that house it are both astounding and terrible at the same time. If these buildings had been constructed with any less care, Allied bombs would have reduced them to a smoking pile of rubble. Even so, the "temporary" roofs installed immediately after the bombs fell were still in place. Some rooms were unheated. There is no modern security system, and no fire sprinklers. If a fire breaks out during working hours, someone is appointed to walk around the buildings banging on a gong. Indeed, the east wing of the museum has yet to be restored, sixty years on, making it one of the few remaining war ruins in all of Berlin.

Labrador Duck 22

Unlike many other treasures in European collections, the Berlin drake narrowly avoided being destroyed by wartime bombing.

After the museum tour, Steinheimer took the Czech researchers on a loop of Berlin, leaving me to complete my usual tricks with the Labrador Duck. A handsome little male, he was a bit dirty but otherwise in good shape. His gray-brown glass eyes had particularly large pupils, making him look more alert than most specimens. Steinheimer had described it as the finest Labrador Duck in the world, but it isn't quite that high up the list. He sits on a simple but elegant wooden base with a slot to hold a Plexiglas cover. The duck normally lives in a glass-fronted cabinet with a stuffed Great Auk and a couple of exotic-looking black birds with long, curvy bills. The tag around his left leg was not particularly revealing, giving only its catalogue number, 14094, and the printed words *Zoolog. Museum Berlin.* Even if we don't know exactly where or when this Labrador Duck was collected, we do know a little about how it came to be in Berlin. Writing in 1954 about extinct and endangered birds in the collection, Erwin Stressmann explained that Martin Heinrich Carl Lichtenstein purchased the specimen for 18 thaler from a Hamburg dealer named G. A. Salmin on August 17, 1838.

To show what a complete prat I can be, Steinheimer had been able to correct me on an assumption. What I had thought was the site of the Berlin Wall the night before was merely a construction site, fully 500 yards inside the region formerly known as East Berlin. He showed me on a map where I could find the river that had made up part of the boundary between East and West, and I set off for it, en route to the train out of town.

On the west side of the Sandkrugbrücke spanning a canal off the Spree, I found a small monument erected to the memory of the first person killed trying to pass from East to West. His name was Günter Liftin and he was twenty-four years old when he tried to make the crossing, just a few days after the border was closed. Günter was on the bridge when he was spotted by guards. They fired a few warning shots, ordering him to return. Perhaps in desperation, perhaps in panic, he jumped into the river. The west bank belonged to West Berlin, but the river itself was part of East Berlin, and the security detail shot Günter in the neck, killing him. He was the first of more than 180 deaths of this sort. When I crossed the bridge a couple of fellow pedestrians stared at me, probably because I was burdened by my yellow-and-black backpack, that makes me look like a giant bee. I was free to cross back and forth over the river as many times as I wished; Günter had tried to cross it once and had lost his life. I took the small monument to be not just a remembrance of his brief life, but also of the complete freedom of movement that eventually came to Berlin. The monument instantly became my lasting impression of the city.

I HAVE NO doubt that Paul Hahn worked doggedly compiling his list of specimens of extinct North American birds in the late 1950s, but he didn't quite get them all. He missed the Labrador Duck in Braunschweig. Goodness knows I wouldn't have found it except by a bit of good luck and the help of a colleague. Today there are online chat groups for just about everything from the relative merits of Roman gods to clothing tips for strippers, including a group for curators of museum bird collections in North America. I used the service to circulate a general inquiry about Labrador Ducks, and received a response from Pam Rasmussen, a zoologist at Michigan State University. She said that while making the rounds of German museums a couple of years earlier, she had taken the opportunity to photograph a Labrador Duck at the Staatliches Naturhistorisches Museum in Braunschweig. She was even good enough to send me the photos, which showed a rather ratty-looking immature male.

Braunschweig isn't a big city, at just 250,000 people, but the *Hauptbahnhof* and zoology museum were at opposite corners of town. Since I had to visit the train station three times and the museum only once, it would have made sense to reserve a hotel room near the former. Instead I had opted for a hotel in the city's far northwest, part of a very large and very economical chain.

Before getting a meal, wanting to be sure that I could arrive at the museum the next day with the least amount of fuss and muss, I decided to go for a wander that evening to find it. But in my quest, I made two fundamental mistakes. The first was to forget my compass. Clouds covered the skies, so I couldn't fix north. The second mistake was to trust a map taken from a tourist guidebook in my hotel. Maps put out by tourist information offices are not designed to get a tourist from one spot in a city to another. They are designed to give the naive traveler the impression that it is as easy as anything to get from one spot to another. I confused north with east, and found myself very, very lost. Using all of my navigation skills, following numbered tram lines and the moss on trees, I managed to find the university, and from there found the museum, and from there found a very nice little restaurant with sidewalk seating. I ordered a *Bier* and a pizza. The beer wasn't good, but the pizza was exceptional. The young university crowd streamed by, some on their way home and some on their way out for the evening.

The next morning saw me at the natural history museum at a fashionable 9:30. In terms of its research collection, the museum is a mid-sized collection with 27,000 stuffed birds, representing almost half of all the bird species in the world. The institution had been founded exactly four hundred years before I got there, the private collection of Herzog Carl I von Braunschweig und Lüneburg. Herzogs—German dukes—seemed very keen on starting natural history museums. This

Herzog's official portrait shows him in military garb, but it appears that he fought most of his campaigns indoors or after dark, somewhere close to a restaurant serving big portions.

I had practiced my little speech in German that ran along the lines of "Hello, my name is . . . Please, I would like to see . . ." I had even rehearsed a little backup in case my contact, Michaela Forthuber, had forgotten about me and gone on vacation. After all, I had made the appointment several months earlier, and had not got a reply to my reminder a week before my journey. Even so, when I got to the reception desk, I did not get past "*Guten Morgen. Mein Name ist Glen Chilton. . . .*" before I was whisked away like visiting royalty and brought to Forthuber.

I gather that Forthuber is the museum's taxidermist, and by default is the curator of some of the collection. She is young and pixie cute, with big eyes and hair cut short enough to indicate that she found life too much fun to waste a lot of time fussing with a styling comb. The high color in her cheeks was probably the result of the warm weather. I tried without success to convince myself that she was blushing because I was so charming. Cute young women don't generally have a lot of time for frumpy old gits like me, particularly since I was wearing the dress shirt that makes me look fat.

Forthuber took me to the bird collection room, and flipped up a little work table attached to the wall. Before settling down to work, I ducked under the table to check that it was properly secured. It wasn't, and so I fixed it. I don't know where the Braunschweig prison is, or what the penalty is for damaging a priceless artifact, but I wasn't keen to find out.

Labrador Duck 23

Well, if not priceless, then certainly valuable. Valuable, but not pretty. If you saw this Labrador Duck at a jumble sale, and didn't recognize it for what it was, you wouldn't pay a dollar for it. It was prepared as a study skin, rather than as a taxidermic mount, and the preparator had not been entirely committed to artistic integrity. Small wonder that the museum's official booklet has a nice photograph of its Great Auk but no mention of its Labrador Duck. It is very probably a male, although a particularly young one. He had a few brown spots turning black, and a few brown spots turning yellow, but was otherwise, well, brown. He isn't a very symmetric specimen, and there is no way to make him lie flat. In order to protect him, he had been placed, rather wisely, by himself in a glass-topped wooden case of the sort normally used for pinned insect specimens. He had even been attached to the case's base with large pins to keep him from rolling around. The tags around the leg of this duck suggested that it had come from the collection of Heinrick (or Heinrich) Ferdinand Möschler, who was born in 1800, died in 1885, and spent part of the intervening period amassing a natural history collection.

As I got toward the end of my work, that silly little voice in my head said that I had missed something. All of the measurements were made . . . all of the drawings had been drawn . . . Forthuber is pixie cute. . . . It took me a while, but it finally dawned on me that the clue I was missing was olfactory. I sank my nose into the belly of the duck and took a good sniff. Most museum specimens smell like mothballs. This one smelled like wood smoke. Definitely not cigarette smoke, or even fireplace smoke, but deep-woods campfire smoke. There were several other stuffed birds on an adjacent bench, and I took a good sniff at those. They correctly smelled of mothballs. I pointed out the duck's smell to Forthuber, who took a tentative sniff. "That isn't preservative?" she asked. No, it's wood smoke. Wait a minute . . . what does arsenic smell like? Almonds? No, that's cyanide. I was getting confused. Is it possible that I had just poisoned myself by sniffing a duck? Oh well.

THE DAY WAS young and mercifully cool, so I headed for the city center, depositing myself in the Burgplatz, the historical and geographic center of town. As noon approached, I expected to be overrun by the booming of bells from a clock tower, but instead heard a gentle tinkling. Heading off to find the source, I found a wedding ceremony that had come to an end at the Rathaus, which I expected to mean "Rat House," but which my phrase book translated as "Town Hall." The happy new couple emerged by running through a paper cut-out of a heart, accompanied by the clapping of hands and a shower of rose petals.

The Marienkirche in Berlin may have been covered in tits, but the *Dom* St. Blasii in Braunschweig was a perch for kestrels, noisy little hawks that strike fear into the hearts of small songbirds. St. Blasii is the patron saint of throats. The story goes that Blasii revived a young lad after he had died when a chicken bone got stuck in his throat. Given that saints are not found on every street corner, this seemed to me another argument in favor of a vegetarian lifestyle. St. Blasii was beheaded in 316, but went on to become one of the most popular saints in the Middle Ages.

I found the Burgplatz nearly free of tourists, and only six were hiding in the *Dom*. They all left as soon as I arrived. This left the lady behind the reception desk and me to keep each other company. She was very concerned about something, but I couldn't quite make out what. She wasn't exactly angry with me, but she wasn't entirely impressed either. There was a lot of gesturing, and she used a loud voice that didn't seem entirely in keeping with a house of worship. Was it my briefcase and duck-detection kit that she didn't like? Did she want me to leave them with her behind her desk? Well thank you very much. *Danke*. It was very kind of her to watch over them for me. She still didn't seem too happy as I walked further into the church.

I checked out a revolving rack of interpretive brochures. I found guides to the church printed in twenty languages, including French, Turkish, Vietnamese, Bulgarian, and Arabic, but nothing in English. The lady behind the reception desk was still eyeing me suspiciously, so I grabbed the first brochure that came to hand. It was written in Norwegian. I sat in silence in a pew, pretended that I could read Norwegian, and enjoyed the cathedral's architecture. It was large and airy, bright and cheery, partially because the windows did not over-emphasize stained glass. A little further along I found a crypt and deposited a €1 coin so that I could visit 24 tombs, including those of Elenore Charlotte (1686–1748), Friedrich Wilhelm (1663–1732), and the other Friedrich Wilhelm (1771–1815). Set a little deeper in the crypt were three more sarcophagi. If I can trust my translation of Norwegian, one of the tombs contains the rather cramped bodies of three persons from the eleventh and twelfth centuries. A second box contains the earthly remains of Heinrich der Löwe, and the third has his wife or girlfriend, Mathilde, an English princess. Der Löwe seemed to have had a finger in every pie in the twelfth century. In 1166 he had a bronze lion erected in the Burgplatz as a symbol of his authority. He commissioned the construction of a castle and ordered the building of St. Blasii's cathedral. Back on level ground, the lady at the reception desk had remembered a smattering of English. In just two words, she encapsulated all that she had wanted to get across when I arrived. She said, "Please leave!"

IF YOU EVER find yourself riding a Greyhound bus through southern Saskatchewan, and the driver stops at every community big enough for a gas station or a Chinese-Canadian restaurant or a church or a pocket-gopher colony, then you are on the "milk run." You will finally dismiss the possibility of getting to your destination before you die of old age, and start wondering if you will get there before the sun goes nova. The expression *milk run* may be quite common in other languages, I suppose, although in Germany it would be described as the "*Milch Weg*." The next day I found myself on the *Milch Weg* out of Braunschweig, headed for Halberstadt.

Forgive yourself if you have never heard of Halberstadt. It just isn't that big, and it isn't mentioned in any guidebook I've come across. To find it at all, I had to shell out big bucks for a 1:750,000 Michelin map of Germany, and consult it with a magnifying lens. Before arriving in Halberstadt, the train stopped at Stapelberg, Ilsenburg, Minsleben, Heudeber-Danstedt, and Wernigerode. Some stops seemed to be providing service only for cell phone towers and wheat fields. At some stations, the train stopped for less than twenty seconds, and the wheat wasn't quick enough to board. As the time for my arrival in Halberstadt approached, I could see several lovely church spires, and I hoped that they marked the city center.

At a tender young age, Ferdinand Heine Sr. (1809–1894) began collecting stuffed birds, and once he got down to it, went at the collecting game at a fever pitch. He must have had more profitable behaviors as well, because he became a wealthy landowner. His natural history collection was turned over to the city of Halberstadt thirteen years after he died and formed the basis for a museum. Of the

18,000-odd stuffed birds in the collection today, fully 11,589 came from Heine. His son, Ferdinand Heine Jr. (1840–1920), got in on the act in a lesser way. He was, presumably, a perfectly nice guy, but when the city named the institution the Museum Heineanum, it was for the father, not the son. Besides a precious Labrador Duck, the museum has six other species of extinct bird: a Passenger Pigeon, a Carolina Parakeet, an Ivory-billed Woodpecker, an Emperor Woodpecker, a Norfolk Island Kaka, and a *Lappenhopf*, whatever that is.

I had been warned that the Halberstadt Labrador Duck might not be quite what I was hoping for. In 1959, Paul Hahn had received a letter from Kuno Handtke of the Halberstadt museum, indicating that some of the skin on the neck of the specimen had come from a Mallard. These things happen, and as long as we know that the specimen is something of a patchwork quilt, no harm is done. However, shortly before my visit, Dr. Bernd Nicolai, current Director of the Museum Heineanum, warned me that the situation wasn't even quite as good as that. Apparently only the head and neck of the specimen are from a Labrador Duck, and the remaining material from specimens of some other species. Well, if the goal is to see absolutely every Labrador Duck in the world, Halberstadt must still be worth a visit, I should think.

Nicolai had written to say that he would be bird-watching in Spain at the time of my visit, but his colleague Rüdiger Holz would be awaiting my arrival. He also told me the museum was immediately beside the city's *Dom*, so I figured navigation by its spires was a good way to get to the museum from the train station. It was another hot day and I tried to avoid the temptation to sweat myself into the ground by walking on the shady side of each street.

Arriving at the museum I was finally able to use my entire "Hello, my name is . . ." speech, and the lady at the front desk had absolutely no trouble understanding me. How odd that for the past ten days, without knowing it, I had been speaking German with a Halberstadt accent. I was guided from the reception building through a gate, across a courtyard, and into an old building to meet Holz.

If you were asked to describe the appearance of a typical ornithologist, you might be inclined to use expressions like *weedy* or *willowy* or *reedy*, or some other plant-related comparison. This is a shame, because most ornithologists I know are hale, hardy, and tanned, wearing denim and leather, and altogether ready to scale the tallest mountain with a machete in their teeth. Holz is not the machete type, despite trying for the look with a rough-and-tumble beard. In describing him, perhaps I can use the word *etiolated* without being insulting, because only botanists know what it means.

Holz started off by showing me the museum's library, which contains volumes on all sorts of natural history topics, with particular emphasis on birds. It has, for instance, a complete run of the *Journal für Ornithologie*. I didn't realize that any library had a complete run. I wasn't shown any books worth €350,000, but the library certainly represents a worthwhile regional resource.

While all very well and good, this was only a precursor to my visit with the Labrador Duck, even if pieces of it weren't genuine manufacturer's replacement parts. My host led me down a steep lane constructed of cobbles that hadn't been reset in a while. We passed a wall with tributes to Halberstadt citizens long dead, which made up part of a churchyard wall. Eventually we arrived at an unlikely-looking unmarked building that Holz explained had been constructed 250 years before as a horse stable. We entered through a locked door, passed through another locked door, and then another locked door, before arriving at the bird collection housed in 120 locked wooden cabinets, stacked floor to ceiling. Halberstadt had not entirely been passed over by Allied bombing, and the museum had lost about 1,000 of its bird specimens. But locked away safe and sound in cabinet 1, along with their Carolina Parakeet and a kaka, was the sixth and last Labrador Duck of my German trip. We cleared some room at a bench, set down the duck, and plugged in a lamp. Holz left me to get to work.

Labrador Duck 24

The specimen in Halberstadt proved to be a fake—a Labrador Duck beak on the body of a crudely painted domestic duck.

All of my warning bells went off immediately. This specimen just didn't ring true. It was too big to be a Labrador Duck, and all of the feathers were either black or white, with none of the gray and brown splotchy patches that I would have expected. Worse than that, it was very clear that all of its black feathers had started life white, and had been turned black with tar-like paint. The black neck ring was too wide. The black mid-crown stripe was too thin and too faint, started too close to the bill, and finished too far back on the neck. The cheek feathers should have been a bit stiff and, in an adult male, should have had a yellow tinge. This specimen had neither attribute. At least the beak was right. It had the right dimensions and the correct flappy bits along the front margin. Sadly, even the bill was messed up, as someone had painted it black and yellow-brown, obscuring some of the finer details.

And so I now proclaim to all and sundry that, other than the bill, the Halberstadt Labrador Duck is a fake. I suspect that somewhere in the depths of time, a lovely Labrador Duck specimen was attacked by moths or mice, such that the only salvageable bit was the bill. Not wanting to throw it away, Heine or one of his cronies had stuck it on the body of a white domestic duck, and painted some bits of it black to more or less resemble an adult drake. Given that the paint job was rather crude, I suspect that the model for the paint job had been a drawing of a Labrador Duck and not another stuffed specimen. And in the couple of hundred years since this was done, no one with enough experience with Labrador Ducks had dropped by to notice the forgery. I suppose that someone could do DNA analysis of feathers taken from different parts of the specimen, but I would wager dimes to doughnuts such an analysis would show that the whole body is that of a white domestic duck. Holz's grasp of English was not perfect, and I didn't want to screw up the explanation, so I didn't tell him. Instead, I dashed off a letter to Nicolai when I got home. I snapped a few photographs, including one of Holz holding the fake.

Before I left, I was invited to sign the museum's guest book. Lots of museums have guest books, but I have never seen the like of the one at the Museum Heineanum. Leather bound and housed in a fancy slip cover, it had signatures of visiting biologists going back to the late nineteenth century. Out of reverence and respect, I used my favorite engraved silver ballpoint pen and printed neatly. Then Holz dashed off to get a digital camera to take a photo of me signing the book. With treatment like this, I was going to start to think that I was somebody important.

WITH FOUR HOURS before my train pulled out of Halberstadt, I was determined to find out what was so reprehensible about the city that it had been expunged from every tour guide to Germany. At the tourist information office, I was given a very good map and an English guide, *Halberstadt and Its Picturesque Surroundings: Your Gateway to the Harz Mountains.* Passing the *Dom*, I had seen that it was the site of some rather serious construction efforts, and I asked the tourist information lady if it was open to visitors. "*Nein*, it is closed to two weeks. They have pets." That's odd, I thought. "Dogs and cats?" I asked. "*Nein*," she replied, "*small* pets." "Puppies and kittens?" I think she may have been going for "pests," which only goes to show you that even God cannot protect you against cockroaches. I also received a small but glossy brochure describing three self-guided walking tours of the town. I had never been on a walking tour, self-guided or otherwise, and over a lunch of salad and beer, I chose to treat myself to Walk Number One, with bits of Walks Two and Three thrown in to flesh out my Halberstadt experience.

My tour started off at the Stadtkirche St. Martini, patron saint of vermouth-based cocktails. The towers to either side of the entrance are of unequal size, and these have come to be an emblem of the town. How odd that a city emblem should be based on a building whose asymmetry probably resulted from a budget overrun. At one end of the *Domplatz*—the cathedral square—stands the Gothic Cathedral of St. Stephen. At the other end stands the Romanesque Liebfrauenkirche. To me it seemed as though the two churches were smirking at each other, each firm in the conviction that its God was bigger than the God of its rival.

In front of the cathedral sits the Lügenstein, the Stone of Lies, or Devil's Stone. According to legend, when Satan saw that a church was being constructed, and not a tavern, he decided to use the rock to destroy the cathedral. The rock is only about 6 feet along its greatest axis, so Satan would have needed a fair few whacks to do much damage. In the nick of time, a tavern was built beside the cathedral, and the Devil was saved the trouble.

Just around the corner, I found number 11 Grubdenberg, the birth house of Ferdinand Heine Sr., founder of the natural history museum's collection. Number 11 is next door to a nice-looking hotel, but the rest of the buildings on the block were boarded up and kind of rubbishy. Down the street, at Bakenstrasse 37, is a complex of buildings apparently known to locals as "little Venice" because a water channel used to flow under it. Or so said my self-guided tour brochure. Nothing suggested this was anything other than a fib, but I was willing to let it go.

Given the nature of my quest, I departed from Walk One, and tromped north to the *Ententeich*, stop number 10 on Walk Two. Now a duck pond, this little ditch was at one time part of the rampart and moat complex outside the town wall. I am very pleased to report that on the afternoon of my visit, there were several dozen ducks on the *Ententeich*. Most of them were snoozing in the shade and the rest were swimming. I didn't have the sense to do either and resumed my tour of Walk One.

Stop number 7 was the *Grauer Hof*, or Gray Courtyard. It is a "charming collection of half-timbered houses, dating from around 1700." It was truly charming, with no line of construction parallel to any other line. I would be a little irritated if someone put my street on an official town tour, but locals sunning themselves in their forecourt were very friendly and waved as I meandered by. The next stop was the *Johannistor*, a town gate and part of the town's fortifications. It had been torn down in the 1800s in order to widen the road. Now, let me get this straight . . . you want me to stop and admire something that was torn down more than a century earlier? There was no picture of the ex-gate, and my imagination isn't that good.

A little further along was the *Johanniskirche*. Constructed 350 years earlier, it is the biggest half-timbered church in Germany, and has a freestanding bell tower. Regrettably, the gate was locked, but being a pretty sneaky sort of fellow, I found a back entrance and was able to admire both the church and its freestanding bell tower.

I finished my tour with Walk One, stop number 13, a sculpture by local artist J. P. Hinz, attached to the side of the telephone exchange building. The work is entitled *Joy of Being Alive*. I was not immediately impressed, so I stepped back into a field of daisies to get a better look. As much as I tried, I couldn't make the sculpture bring forth a joyous feeling. To me, it sort of looked like three musicians being crucified, while two other people lounged nearby, pretending not to notice. However, the House Sparrows, nesting in the sculpture's various nooks and crannies, made me feel joyous.

So what can I say about Halberstadt? The self-guided walking tours are a little goofy, but that isn't necessarily a bad thing. Not every retail space is occupied and not every patch of grass is uniformly green, but everyone I passed seemed pleasant and happy. The trams run nearly silently, and the buses are really posh, having been built by Mercedes-Benz. The fountains all had water and the town has a rich choice of dining opportunities. All in all, I think that editors of guidebooks to Germany need to give Halberstadt another look. My tour of Germany was coming to an end. It had all come down to a train ride, another train ride, a long walk, a taxi ride, another train, another train, a bus, a plane, another train, and one last train to take me back to Lisa. None too soon.

Chapter Twelve

A Black-and-White Duck in a Colorless Land

Are you up for a little challenge? It may be trickier than it first sounds. Try to name five famous Belgians. Male or female, ancient or contemporary, rich or poor. I'll even spot you the muscled actor Jean-Claude Van Damme, so you only need four more. I was once told that the most famous Belgian of all is Tintin, but since the boy reporter is, of course, a cartoon character, he can't be counted as one of your four, and Agatha Christie's fictional detective Hercule Poirot doesn't count either.

And therein lies one of the problems for poor old Belgium. It just doesn't get the rave reviews showered on all of its neighbors. Most folks can come up with an image of France even if they have never been there. The same must be true for Germany and the Netherlands. But unless you have actually been there, your image of Belgium is probably a bit vague, involving an amalgam of impressions of France, Germany, and the Netherlands. Indeed, Brussels has a reputation for being unassertive, colorless, and just plain boring. In essence, the city has real identity problems. So when I set off to see a Labrador Duck at the Institut royal des Sciences naturelles de Belgique in the capital city, I decided to put a spin on it by looking for a little color. Literally.

This was to be my shortest duck-related adventure. Just a quick hop to Brussels, have a little look at their duck, and hop back out again. My mother, Kathleen, who has as sharp an eye for color as anyone else, joined me for this journey. She seemed very keen to be on this adventure, particularly since she didn't have to do any of the planning, or fiddling around with airlines schedules or hotel reservations. My mother's job was to tag along and have fun.

Traveling with the cheap blue-and-yellow airline, we didn't fly into the real Brussels airport, but rather into the "Brussels South" airport near the city of Charleroi, about one-third of the country away from my target. As our flight descended through drizzle and heavy low clouds, the three predominant colors were blue-gray, green-gray, and glaucous.

After settling into our hotel, we strolled through the early evening streets of Charleroi in search of a meal. It was mid-September, and the sun had long since abandoned its attempts to push aside the rain clouds. The streets were quiet, even for a gloomy Monday evening. Dominating the neighborhood were banks, optical dispensers, employment agencies, and prostheses shops, punctuated by restaurants and bars. A few businesses displayed neon lights in their advertisements, which were reflected on the rain-slicked streets. The evening seemed to cry out for the strains of a saxophone or an accordion. You may choose to add Adolphe Sax to your list of famous Belgians. Adolphe invented, and named, the saxophone. I have no idea who invented the accordion. We found a bright and warm restaurant that served food of many types. With the help of my French-English dictionary, I translated the menu. My mother chose a ham and cheese (*jambon et fromage*) concoction from the "small appetite" column. I had *penne avec quatre fromages*. We washed these down with surprisingly small glasses of Bass Pale Ale. Back at the hotel, Mom got into the spirit of the color-themed adventure by pointing out that the bathroom was decorated in two shades of dull blue.

WHEN I HAD first peeked out of the hotel room window the next morning, the weather had been gray, windy, damp, and cold. On our way to the train station, it had changed to bright blue, windy, damp, and cold. Charleroi isn't a big city, but I still managed to get us a bit turned around and wound up in the red-light district. Only one lady was working the street, and another, nearly naked, lounged provocatively in a chair in a shop window. Not a lot of choice, but it was only 9:30 in the morning.

At the *gare* I went through my prattle in limited French to the man behind the counter. "*Bonjour. Je voudrais deux billets aller-retour pour Bruxelles Centrale, s'il vous plaît.*" I must have butchered the part about return tickets because he asked, in perfect English, "You want to come back?" Mom chuckled, and suggested that I start every conversation with "*Parlez-vous anglais?*" before annihilating the French language.

The clickety-clack train had gray seats, a gray floor, and gray-and-orange walls. As we pulled through the outskirts of Charleroi we passed gray and brown warehouses. Might we find the rest of Belgium as colorless as I had been warned? Some of the villages along the route were small enough that all the homes were detached. In larger towns, folks lived in very tall, very skinny, charming row houses. The train zipped past the town of Waterloo. Several days passed before it occurred to me that this was *the* Waterloo.

After detraining at Brussels' *Gare Centrale,* I told Mom about my unique form of navigation in foreign cities—following people who look as if they know where they are going and hoping for the best. For a change, it worked. We soon found ourselves in the shadow of the spire marking the Grande Place, the center of Brussels in almost every sense. The cobblestoned marketplace has been a gathering place for traders for about one thousand years. Today the architecture of many great European cities is dictated by the reconstruction efforts following World War II bombing. In contrast, the Grande Place owes its character to buildings erected after two days of cannon fire by the French in 1695, but I never discovered what residents of Brussels had done to irritate the neighbors. Trading guilds had constructed their guild houses to match dictates of the city, resulting in buildings in glorious harmony. The judicious use of gold relieves the possible monotony of the gray cobbles and gray-and-tan buildings. Statues of what I took to be kings, knights, saints, and gargoyles festoon the town hall. My favorite was a gargoyle parrot. Some of the statuary may be missing an arm here or a head there, but the overall effect is stunning. The plaza, probably overflowing with sellers and buyers on a sunny day in July, was just pleasantly occupied on our mid-September midweek morning.

Tucked away in a corner of the square, we found the brass statue of Everard 't Serclaes. Mom rubbed his arm, which is said to bring good luck. Not the luckiest fellow himself, he was murdered while defending Brussels in the fourteenth century. From the shine on Everard, he must be rubbed almost continuously. All over. Lucky devil. Before leaving the square, Mom found a shop that sold her a handmade Belgian lace table piece that cost almost exactly as much as our airfare. We then stumbled across one of the most famous landmarks in Brussels. It is, sadly, a small statue of a boy urinating. The original dates to the early seventeenth century. After attempts by the French and English to steal it a century later, it was finally nicked, broken, replicated, and remounted in 1817. In a corner shop close by, I spied the ugliest souvenir on the face of the Earth. It was a two-inch-tall replica of the peeing boy with a bottle opener sprouting from the top of his head.

We walked uphill to the escarpment that divides the two portions of Brussels and entered Parc de Bruxelles, once a hunting estate for mucky-mucks, redesigned as a public park in the late 1700s. We sat on a bench near a fountain at the park's north end, admiring the tree-lined avenues and watching runners fill their lunch hour with fresh air. The sun played with clouds. When the sun briefly took the upper hand, it created a lovely rainbow in the fountain's mist. Then the wind changed direction, giving us a good hosing down.

From the park, we walked east down the rue Belliard, aiming for the European Parliament, on the margin of the Parc Léopold. Just beyond, we should find the Institut royal des Sciences naturelles and their Labrador Duck. Our walk took us through the Big Business and Administrative and Don't-You-Damned-Well-Forget-It District. I expected the odd restaurant for the lunchtime crowd. I expected the odd bar for the after-work crowd. Not a one to be seen. Our lunch would have to wait until after my examination of the duck. Walking through the Parc Léopold and past the Bibliothèque Solvay, we came across a field full of young women playing baseball, using a tennis racquet instead of a bat, and hula hoops for bases. It seemed to be some sort of team-building exercise. Instead of the flamboyant colors of youth, each wore a white T-shirt and black sweat suit bottoms.

According to my contact at the institute, Georges Lenglet, the best way to find the museum building was first to find the European Parliament buildings, and then look for a yellow-and-orange tower. The first part was easy. The European Parliament is a behemoth. Then we scanned for a yellow-and-orange tower. Well, I suppose that color might be called yellow on a really sunny day, and that color could be mistaken for orange if it had an immediate transfusion of red. And yellow.

The Institut royal des Sciences naturelles is not the oldest museum in Europe. It was inaugurated by Léopold II in 1891. The builders designed it to show off a particularly stunning set of iguanodons found near Mons twenty years earlier. The museum's promotional material didn't explain what the iguanodons were doing in Mons. The museum's ornithological collection ranks twenty-fourth in the world, with about 100,000 items. The portions of the museum open to the public display lots of minerals, shells, and stuffed animals and their skeletons, but the real highlight is the dinosaur exhibit.

Arriving at the appointed hour of 13:00, I found that Lenglet was the quietest and least-assuming man I had met on my duck quest. He wore a sensible white lab coat and held his white-crowned head bent forward slightly, as though his brain were too heavy to be held perfectly upright. Anyone whose job title includes expressions like *vertebrate systematics* and *biochemical taxonomy* surely has a rather full brain. He was gentle, polite, and helpful, and probably had a lot of good stories to tell, but I bet those stories aren't drawn out of him easily. He passed completely on the usual pleasantries of: "How was your journey? How are you enjoying Belgium? Did you manage to find the Charleroi red-light district? Did you like our statue of the peeing boy?"

Lenglet took us to room 15.55, then rolled the Labrador Duck down the hallway on an antique lab cart to his office, where I was to complete my examination. I appreciated his exaggerated care in negotiating every bump in the floor and every turn in the hallway. Not everyone treats these specimens with sufficient reverence. He pulled up a chair for my mother while I examined and measured the duck. He also earned very big points from me by fetching her a cup of coffee.

Labrador Duck 25

This was another taxidermic preparation of an adult drake. While making detailed notes about the appearance of the duck, a strange question came to mind. If, years from now, a Labrador Duck were stolen, and then one came up for sale, would my notes be sufficiently detailed to give evidence in a courtroom? This one was a pretty standard-issue Labrador Duck, but I felt it was notable for two things. First, it was remarkably clean for a very old specimen. You can put this down to the glass case that fits over the preparation. Second, his tail was broken. Not as in broken off, but as in split into a left and a right side. I wanted to have a really good poke to figure out what might have happened to it, but felt that if I did, it might drop off in my hand. I also had notes about exactly where it was painted (legs, toes, and webs, but not the bill) and with what colors (black). I jotted in my book that it had small holes in the inner web of the left foot. I measured the bill and wings with precision. So, if a defense attorney ever asks me, "Are you sure this is the same duck you saw in Brussels forty years ago?" I will be able to give my answer with certainty. By then, stealing an extinct duck will probably be a capital offense. Regrettably, virtually nothing is known about the origins of this specimen. The Royal Institute got it from another Brussels museum sometime before 1845. With a sense of humor that not everyone would understand, my mother named this duck Georges, after our host.

Giving my eyes periodic breaks from squinting at the duck, I looked around Lenglet's office. It was an absolute masterpiece. At least eight times as big as my office, it had a wealth of bench space, plenty of bookshelves, and room for all the filing cabinets that a scientist could ever want. The office was many floors up, and with windows on two sides, it had a great view. There were stuffed fish on the wall and foul-looking creatures in jars of foul-looking preservative. To add to the whole effect, the office had a spiral staircase that led to a second level, which housed Lenglet's research library. Tucked in the corner of the library was a grotesque skeleton. It was like a road accident, and I couldn't help but look at it. Brown with age, it was the skeleton of a very small child, topped by a huge and grossly deformed skull. "Hydrocephaly?" I asked. "Yes," Lenglet responded. He described his office as a museum within a museum.

LEAVING THE INSTITUTE, Mom and I zigzagged through the buildings of the European Parliament before aiming for the train station. Don't let anyone tell you that Brussels is all gray. Some of it is silver, and some of it is sand-colored. I explained to Mom that I particularly hate tall glass buildings because of their impact on migrating songbirds. Traveling at night and navigating by the stars, when they pass over cities they can become disoriented by twinkling lights reflecting off the glass. Many die when they collide with buildings. To illustrate, I pointed out a small dead bird at the edge of the sidewalk.

We descended to train platform 6 to await the 15:36 train to Charleroi with about one hundred other travelers. At 15:30 a train pulled up to the platform. An information screen confidently told us that this was the 15:36 to Charleroi. A few people got on board, but most of us held our ground, looking at each other skeptically. One minute later when the doors closed and the train rattled away, the information screen told us that we had all missed our chance, and that the next train would depart at 15:47 for Brussels Midi. Everyone groaned, and a few vulgar words were uttered in both French and Flemish. "Oops, no, wait a minute," proclaimed the information screen. "My mistake," it said. "The *next* train will be the one for Charleroi. Sorry!" I'm sure that I wasn't the only one who wondered where the riders on the 15:30 mystery train finished up.

BACK AT THE hotel in Charleroi, I worked on getting down on paper my thoughts on the day while Mom had a quick nap. She awoke keen to go in search of a beer, having earlier spotted an upmarket bar done over in brass and wood and subtle lighting. We ordered a couple of Leffe, which arrived along with a little glass bowl of cheese cubes adorned with garlic salt. I told Mom a few stories about ducks. She told me a few stories about our relatives and other rogues. I pulled out the euro coins that had accumulated in my pocket since arriving in Belgium. In the currency of the European Union, bills are the same wherever you are, but the coins of each country are unique. Despite the differences on the "heads" side, all coins are legal tender in all countries in the union. As evidence of how truly European Belgium is, I had accumulated one coin each from Ireland, Germany, Spain, and France, four from the Netherlands, and two from Belgium.

I suppose we should have gone in search of food, but with a couple of beers and a couple of bowls of salty cheese in us, a long tramp seemed to be more important. In the remaining light of an early evening we wandered past shops that, at an earlier hour, would have been pleased to provide eyeglasses or glass eyes, women's dresses or undressed women. A couple of times Mom said, "Let's just walk to the end of this block, and then we can head back before my legs wear out." But at the end of each block, she would spot a fountain or a clock tower or a series of particularly tall, thin homes, and we would walk some more. Pausing for a break on a park bench along a tree-lined avenue, she said: "When the time comes, I want you to say my eulogy. Remember to tell people that I knew how to laugh, and that I wouldn't have done anything different."

Then she got a hankering for a banana. We found a number of greengrocers that were still open. The first two had no bananas. The third had a boxful, but they were all brown. The fourth had bananas galore, and they seemed to be just ripe enough. "Are you going to go in and get them?" she asked. Under different circumstances I suppose I would have, but I wanted to see how she would get along speaking beer-fueled French, and handed her a fist full of coins. She came out a few minutes later with a big bagful. "'*Trois,*' I said. He said, '*Kilo,*' I said, '*Non, trois.*'" They settled on a kilo for one and a half euros. "That's all right, is it?" she asked as she handed me the remaining change. "Yes," I said. "That's just fine."

Chapter Thirteen

Echoes of Once-Great Voices

I really thought that I had nailed it. Surely I had located absolutely every single stuffed Labrador Duck in the world. I had perused Paul Hahn's 1963 inventory until my eyes were full of gravel, contacted most of the curators of the world's bird collections, and followed every insidious tendril of the World Wide Web. Fifty-two stuffed Labrador Ducks and one beak had survived the ravages of time, right?

But then Frank Steinheimer sent me a message from Berlin explaining that he had discovered another duck, residing in the teeny-tiny village of La Châtre in the middle of France. Blast! There was even a photograph of this specimen on the website of the village's Musée George Sand et la Vallée Noire. From what detail I could make out, it seemed real enough. If I had known about the duck a year earlier, I could have snagged it while in France with Julie. Now to ensure that no Labrador Duck would escape my quest, I would have to make a return journey to France and revise my global list.

Oddly, the museum in La Châtre was not dedicated to extinct ducks, but to the nineteenth-century French author George Sand. Sand is a really big name to those who care about period feminist literature, and there is probably no end of biographies. However, my branch of the public library had only one, a 1999 tome entitled

George Sand: A Woman's Life Writ Large, by Belinda Jack. The book is really long. Really, really long. It has 411 pages before the explanatory notes. Even having taken six credits of university English literature, I found Jack's writing opaque. As evidence I offer up the line: "And the physical freedom of her subversive capering in La Châtre not only provided material for prose accounts, it also served as a metaphor for more fundamental shifts in consciousness." I didn't finish the book. Nonetheless, I did manage to pan a few nuggets before giving up. Here are ten things to know about Sand in case you are ever called upon to write a snap quiz.

1. Born in Paris on July 1, 1804, she was christened Amantine-Aurore-Lucile Dupin. She picked up the handle George Sand later in life. To family and friends she was Aurore.
2. Sand's father was Maurice François Dupin, an army captain. Her mother was Antoinette Sophie-Victoire Delaborde, a mentally unstable former prostitute. In no big hurry, Maurice and Antoinette waited until four weeks before Sand's birth to get married. She was not exactly born out of wedlock, but certainly not very far into it. Suddenly finding themselves in a big rush, they had Aurore baptized the day after her birth.
3. Dupin married Delaborde, who was well below his station, in secret, and against his mother's wishes. The matriarch eventually came around and arranged for Sand's First Communion at La Châtre on March 23, 1817.
4. Snooty, perhaps, but no purebred herself, Sand's grandmother was the illegitimate daughter of Maurice de Saxe. A step further back, we find that de Saxe was the illegitimate son of King Augustus III of Poland, and his mistress Aurore de Kœnigsmark. De Saxe's niece married a son of Louis XV. After watching all the leaves fall off the family tree, this means that Sand was second cousin to Louis XVI, Louis XVIII, and Charles X.
5. Sand's stuck-up grandmother recognized that Paris wasn't the safest place to be during the French Revolution, and so took up residence in Nohant, just down the road from La Châtre.

6. Sand married Casimir Dudevant in 1822. They had a son and a daughter, but their paternity isn't crystal clear. This leaves me wondering if a big sexual appetite and a propensity toward infidelity run in the family.
7. Biographer Jack described Sand as a "frigid, bisexual nymphomaniac." To me, two of those words don't go together. She was also a cross-dresser, and smoking was one of her nasty habits in a time when women just didn't do that sort of thing.
8. Sand's books were a commercial success, partly because her writing was easy to read, and partly because it was smutty, at least by the standards of the day. Amongst her favorite themes were sex, sexuality, incest, infidelity, sex, and the role of surrealism in contemporary art. All right—I made up the last one.
9. Keen on spreading a little joy, Sand took her share of lovers, and the share of a few others as well. Her most famous consort was the composer Chopin. Among her "friends" were musician Franz Liszt, authors Flaubert, Balzac, Baudelaire, Dostoevsky, and both Brownings, along with a small but elite army of visual artists. Of her lovers, Sand seems to have saved an awful lot of energy for her relationship with Marie Dorval, a famous beauty of the Paris stage.
10. Not necessarily a ray of sunshine, Sand described life as "a great wound that never heals." I think I dated her once. Sand died in Nohant in 1876.

GETTING TO LA Châtre to see its duck was no easy matter. The small town is smack dab in the middle of France, and not close to anything else. Getting to the nearest airport, Limoges, required a five-and-a-half-hour layover at Stanstead. I spent much of that time swotting up on George Sand. She wrote great piles of books, including novels and plays, essays, a two-volume biography, and twenty-five volumes of correspondence. Shelves of university libraries sag under her literary output, and students of feminist literature probably wish she had dedicated a little more time to bonking famous composers and a little less time to writing. Her first novel, written when she was twenty-seven, was entitled *Indiana.* I purchased a copy, and over the next few days, I found spare moments to read it on trains, in airport departure lounges, and on park benches. *Indiana* reads as though written by a thirteen-year-old girl in love with a pony.

The Limoges taxi lobby must be really strong, for there was no bus service into town, and the cost of the short journey from the airport was scandalous. After a short nap in my hotel room, I made my way to the Gare des Bénédictins. If there were only a couple of trains between Limoges and La Châtre the next day, I wanted to make sure I caught one. I asked the only free agent at the information kiosk if he spoke English. "*Non,*" he said, pointing to his colleague at the next spot. When my turn came I asked, "*Parlez-vous anglais?*" She dropped her face into her hands, perhaps hoping that I would go away. So, trying to be polite, I did my best in French, explaining that I needed a return ticket to La Châtre for the following day. I left the last consonant off "Châtre" as my guidebook instructed. She looked at me as though I had asked for a return ticket to her grandmother's underpants. I tried again, sticking the last consonant back on. She sighed and pushed a pencil and a piece of paper at me. I wrote "La Châtre" and pushed them back. Three of her colleagues wandered over to join her, apparently under the impression that this was the height of entertainment. "Look," I said in English, "it's here on the map" pointing to a spot just 55 miles from Limoges. All four of them started chuckling at me. At me, but definitely not with me. With my map as proof, they had to admit that La Châtre actually exists. In French, one of them asked me why anyone would want to go to La Châtre. "*Je voyage là pour les bon marché drogues récréationnelles,*" I said. They convinced their computer to spit out a timetable. It showed that the train doesn't go to La Châtre. The number 3630 train would require sixty-three minutes to take me as far as Chateauroux, a huge overshoot. I would then have eight minutes to use *un autre mode de transport* to get to the *gare routière*, and hop on the number 42693 bus to La Châtre. Three hours and eighteen minutes after I arrived, I would have to be on the number 42706 bus back out of La Châtre, unless I wanted to spend the night on a park bench. "*Billet ici?*" I asked. "*Non,*" I was told, and in the fashion of

Saint Michael directing Adam and Eve out of Eden, each of the four employees pointed to a distant kiosk. The lady who sold me the ticket was far more pleasant, but crushed my ego by asking if I was eligible for a senior citizen's discount.

THE NEXT MORNING, by train and by bus, I aimed for the museum named in honor of George Sand. I arrived two hundred years, three months, and twelve days after her birth. Sometime in the fifteenth century, the Chauvigny family built a castle in La Châtre. Time saw most of the castle fall down, as time so often does, but its stone tower remains. Sixty-five feet tall, with walls six feet thick, the tower is visible from most parts of town. From 1734 it served as a prison, but in 1937 it was converted to a museum to house Sand memorabilia and, two years later, the town's collection of birds.

I was met by Brigitte Massonneau, who had been corresponding with me by postcard. She pleased me by speaking not a single solitary word of English. I knew that I could count on my elementary grasp of French if the person I was speaking to knew an equal amount of English. My interactions with Massonneau showed me that I could stumble along even if my French companion was completely unilingual. As she fetched a table and chair from a storage closet, I looked at the cabinet, just down from the reception desk, that housed the duck. He looked real, and I was more than a little relieved. It would have been a long trip for another painted domestic duck. The cabinet also housed a Carolina Parakeet, an Ivory-billed Woodpecker, an Eskimo Curlew, and a Passenger Pigeon. Far and away the highlight of their collection was the Labrador Duck.

The museum has something like 400 stuffed birds on display, part of its collection of 3,000. The collection came together in the eighteenth century under Jean-François-Emmanuel Baillon, a lawyer and bailiff in the town of Abbeville, very close to the English Channel in Somme. Whenever he got a few free minutes, he spent them in the world of natural history, collecting and stuffing birds and corresponding with great naturalists of the day. His son, Louis-Antoine-François Baillon, picked up his father's passion for all things outdoorsy. When he was twenty years old, the younger Baillon moved to Paris to take up the post of assistant naturalist in the botanical garden. After just a few years, the elder Baillon snuffed it, and his son returned to Abbeville. He expanded on his father's collection, bringing it to 6,000 specimens. When the younger Baillon died, in 1855, the collection was divided in half. The first 3,000 birds went to his daughter, a Mrs. Delf, who stored them in a wet cellar until they rotted. The second 3,000 birds had a better time of it. They were given to Baillon's other daughter, the wife of Philippe Bernard, a physician in La Châtre. Philippe Léonce, a Major General and Grand Officer of the Legion of Honour, inherited the surviving specimens and turned them over to the Village of La Châtre in 1888.

Labrador Duck 26

It was now 11:00. The museum was due to close for a two-hour lunch at noon, and, promising Massonneau to be finished by then, I tucked in. The duck, a male stuffed by a taxidermist, looks as though he has had a rough time of it. He has no glass eye on the right side. His left eye, quite dirty and possibly brown, is sunken in its socket. The webs of his feet and some of his toes have been nibbled away by mice. His tail is bashed up, probably post-mortem. Overall, he is a bit scruffy. Despite all this, I felt excited to be in the presence of a "new" Labrador Duck.

According to an assortment of labels and plaques associated with the specimen: "*Canard du Labrador*: (Camptorhynchus Labradorius), Anatidae, un mâle du Haut Missouri, capturé par Wied (USA), espècies éteinte depuis 1875." I also found that "La ville de La Châtre" had been given the specimen by "Mr. LE Général de Division L. de Beaufort."

If these labels are anything to go by, the drake in La Châtre was collected by German naturalist Prince Maximilian zu Wied-Neuwied while traveling in North America, perhaps somewhere near the upper reaches of the Missouri River, or upper Missouri State. This is the same von Wied who had collected the pair of Labrador Ducks now in Leiden and one of the adult drakes in the collection of the American Museum of Natural History in New York.

In the hour I spent poring over the duck, the museum didn't receive any other visitors. The telephone rang twice, and both times I heard Massonneau say, "*Canard du Labrador*," and "Canada" and "Professeur Chilton." My name sounds a lot sexier when said with a French accent.

AND ALL THIS left me with nothing to add but a long string of thank-yous to Massonneau for her help and a two-hour stroll around town before my bus left. Luckily, La Châtre is the sort of town that can be strolled in two hours. Home to 4,700 inhabitants, La Châtre is renowned for its annual stone sculpture competition. A stage of the Tour de France runs through town. And . . . after that, it's pretty much down to George Sand. Two hours wouldn't leave me with time to get to the *cimetière*, the *gendarmerie,* or the *hôpital psychiatrique*, but I would get to almost everything else. I started by picking my way down to the river Indre, which splits and melds many times as it flows through town. The river had been used to soak animal hides as part of the tanning process. Two hundred years ago La Châtre was a center of the tanning industry, which died away as tanneries closed one after another.

The tanneries may be long gone, but at least La Châtre had the reputation of George Sand to fall back on. If someone had left me holding a dead cat, I would have found it difficult to swing it in La Châtre without hitting something linked to George Sand. I stopped to admire the Maison de Bois, Place Laisnel de la Salle. In one of Sand's novels, the heroine, a florist, lives in the house. I passed the Maurice Sand Théâtre, named after George's son, a theater producer and puppeteer. Slipping behind the town hall, I came across the George Sand Junior High School, then stumbled across a statue of George Sand, commissioned the year after she died. This seemed an odd tribute from a town Sand was said to have despised. The statue is flanked by two 114-year-old giant sequoia trees, representing Sand's position in nineteenth-century French literature. Or perhaps someone on the town council just liked sequoias. I walked along George Sand Avenue, peered down Impasse George Sand, and discovered the Lycée George Sand.

Without information to the contrary, I assumed the Église Saint-Germain is the church in which Sand's grandmother arranged for her to take her First Communion. Bits and pieces of the church date back to the eleventh century. Money was donated for reconstruction in 1895, but due to miscalculations about the strength of the foundations, the church came tumbling down the following year. It took eight years to stick it all back up. I then took in the Maison rue des Pavillons, where Sand took up residence after separating from her husband.

I found myself in the Place du Marché. As far back as the 1400s, this was La Châtre's main square and the very heart of its industrial and commercial workings. At one time the Place du Marché was the place to be big and important and cool. Sand purchased wax and tobacco there. If I had wanted either, I would have been out of luck; like the rest of the village, the shop was closed for a two-hour lunch break.

I removed myself to a smaller square known as Place du Docteur Vergne, after someone who performed cataract operations, where George Sand's parents used to smooch covertly. Perhaps Sand was conceived here. My wanderings led me to a lovely little grassy square known as the Place de l'Abbaye. Two centuries earlier it was a little less manicured than it is today, and reports say George Sand frequented the square with her botanist friend Jules Néraud, to whom the concluding chapter of *Indiana* is dedicated. Unable to find an open rest room, I utilized an isolated spot behind a wall. Next to the square, I stumbled across the house of François Ajasson de Grandsagne, one-time mayor of La Châtre and father of Stéphane, who taught George Sand biology, and probably bonked her silly.

In my two-hour walk through La Châtre, I spied only one other tourist, working off the same town map I had picked up at the tourist information office. Like me, she stopped at the village-approved highlights, read the blurbs provided, smiled, and walked on. She was carrying a copy of *Indiana*.

Chapter Fourteen

Death in a Viennese Sewer

Even though I was slowly draining our bank account, Lisa had been very good about waving good-bye when I took off on trips to exotic spots to see duck corpses. But when it came time to see two pairs of ducks in Austria and the Czech Republic, there had been some pretty good hint dropping about who my next travel companion should be. Being a sneaky sort of person, I somehow managed to turn all this around to come off as the good husband who was taking his wife on a romantic second honeymoon to exotic destinations in Europe. First stop Vienna, then on to Prague.

Before departing, I reviewed what I knew about Vienna. My impressions all revolved around some very large sewers, burly men in army jeeps, a Ferris wheel, and a bit of gunplay, all in black-and-white. My limited knowledge of the Austrian capital all came down to the 1950 film based on the novella *The Third Man*, by Graham Greene. Any sensible married man in Vienna would try to show his wife a good time by visiting sites of great romance. Indeed, the brochures sent to us by the Vienna tourism offices featured an awful lot of photographs of young couples kissing. Instead of being sensible, in addition to examining two Labrador Ducks at the Naturhistorisches

Museum, Lisa and I set out to visit some of the Viennese locales featured in *The Third Man*.

A series of flights brought us into Vienna close to midnight, but, ever the intrepid adventurers, we set off very early the next morning, east along the Mariahilferstrasse, toward the grand center of Vienna. On that chilly, gray October morning, the street was a cliché about the power of retail shopping. If you can't buy it on Mariahilferstrasse, then you probably don't need it. The expanse of pavement devoted to pedestrian traffic was far greater than the corridor for vehicles. All around was a parade of temples to name-brand watches and exclusive clothing. In stark contrast, an impressive series of street vendors were selling roasted *Maroni* from the top of fire-filled oil drums. *Maroni* look a lot like chestnuts. We were invited to sample twelve for two euros, or we could be cheap and have just eight for €1.50. I have never eaten *Maroni,* and they probably taste great, but I was put off by their smell of wet socks.

The retail corridor gave way to Museum Quarter. The Natural History Museum is a grand affair, across the Maria-Theresien Platz from the Kunsthistorisches Museum. Opened in 1889, its displays run the range of natural history broadly defined, from archaeology and anthropology through zoology and paleontology. Entering the museum through the opulent front doors would make one feel part of something grand. We entered through the considerably less grandiose staff entrance off the Burgring. The *Kurator* of the *Vogelsammlung*, Ernst Bauernfeind, was away on the day of our visit. In his place we were greeted by Anita Gamauf, whom I had met at the conference in Leiden the previous fall. Gamauf, a researcher and curatorial assistant, is no giant of a woman. If she were standing on a telephone directory she might be described as medium height, but only by someone not paying attention. But what Gamauf is missing in altitudinal accomplishment she makes up for in enthusiasm.

Usually I have no trouble concentrating when I get down to work, but the behind-the-scenes workspace of Vienna's Natural History Museum is an opportunity for distraction. Now a storage and work area, the room had once been a gallery open to the public. The great arches still sported paintings of cherubs, peering down while I worked. They made me shudder. Lisa speculated on whether creatures are considered cherubs if they don't have wings. Hidden speakers serenaded us with strains of Viennese waltzes, and it occurred to me that it was the first time I had heard a Viennese waltz since Lisa and I took ballroom dance lessons fifteen years before.

Labrador Ducks 27 and 28

With her mouth slightly agape, the Labrador Duck hen looked as though she were talking. With his head cocked a bit to one side, his bill slightly twisted, her drake looked as though he were dutifully pretending to listen. She had been given bright lemon-yellow eyes. His are brown. They rest on wooden bases that are identical, except that the hen's is 2 cm shorter and narrower. Small cards glued to the bases read *Camptolaemus labradorius*, an archaic scientific name for the Labrador Duck, and the relative position of these cards suggested that when these specimens had been on public display, the hen had been situated to the left of the male. The hen had a substantial patch of lighter-colored feathers on her upper neck that I had not seen in other females, but if this signified something important, I couldn't figure out what it was.

The hen was acquired from a well-known Hamburg natural history merchant named Brandt in 1846; her earlier whereabouts are unknown. The drake dates back a further sixteen years, and is one of 160 birds the museum obtained between 1821 and 1832 from Baron von Lederer, who had been swanning around in America, acting as the New York consulary agent for an Austrian blue blood, possibly Emperor Franz I. This duck was probably swept to the great beyond while wintering in or near New York City.

After completing my peek and poke, Lisa retrieved Gamauf, who took us through a series of vault doors, down to the museum's sub-subbasement, where the bulk of the research collection was kept in cold storage. She showed us some of the museum's great treasures, including a really fine Great Auk and some mid-eighteenth-century stuffed hunting falcons, complete with hoods and jesses. The falcons shouldn't have been a surprise, given that many of the early treasures came to the museum from Austria's royal family. Gamauf told us about one important contributor to the museum's holdings, heir to the throne Archduke Rudolf. In 1889, the same year the natural history museum opened, the archduke and his mistress, Mary Vetsera, did away with themselves.

With no particular plans for lunch, Gamauf agreed to join us, leading us to a nice place just across Bellariastrasse. I was a little frightened by the waiters dressed in tuxedos, but discovered that this was the normal situation in Vienna. Lisa and Gamauf had cranberries and fried cheese, and I had a pastry stuffed with spinach. When I asked for a beer, our waiter was far too polite to ask me what sort of beer I wanted, and just used his judgment. Gamauf paid for lunch, and wouldn't even entertain discussion on the matter.

WITH TWO MORE ducks behind us, Lisa and I set off in the tracks of Graham Greene's imagination. I had first read *The Third Man* more than twenty years earlier, and felt I was ready for a reread. The protagonist, Rollo Martins, wrote novels of questionable merit about the American Wild West under the pseudonym Buck Dexter. Martins is invited to Vienna by his school chum Harry Lime. Set in 1950, Vienna is attempting to rebuild itself from the ravages of war. Arriving in the Austrian capital and finding that Lime has died in an automobile accident, Martins makes his way to the cemetery just in time for the funeral. There he meets Colonel Calloway, a British military police officer, who informs Martins that Lime had been involved in deadly black-market dealings in penicillin. Calloway strongly suggests that Martins return to London forthwith. Instead Martins sets about to discover the truth behind the death of his friend. He becomes romantically involved with Lime's girlfriend, an actress of marginal talent. Martins eventually discovers that Lime's death was faked, a ruse to end the criminal investigation that would have landed him in jail. In an attempt to bring Lime to justice, Martins falls in with Colonel Calloway and winds up fatally shooting Lime in Vienna's sewer system. Pretty bleak. Not a chuckle in sight.

Tramping in the footsteps of Rollo Martins, we found the weather very cooperative. It was a bleak February when Martins landed in Vienna. Ours was a bleak October day, like an incomplete eclipse with the sun hidden by low and heavy clouds. At least it wasn't snowing on us as it had been on Martins. We started our tour with a visit to the Hofburg complex and the Neue Burg, not far from the natural history museum. On page 68 of my copy of *The Third Man*, Martins is stuffed into a car and believes that he is being driven somewhere secluded to be murdered. From the racing car, Martins "caught sight of the Titans of Hofburg balancing great globes of snow above their heads, and then they plunged into ill-lit streets beyond, where he lost all sense of direction."

The Hofburg complex has been the center of Viennese administration for six hundred years. Once housing the imperial apartments, it now houses the offices of Austria's president. In Greene's day, the region around the Hofburg would have been swarming with Russian, French, British, and American army personnel sporting rifles. As we arrived at the plaza in front of the Neue Burg, we found it swarming with Austrian army personnel sporting rifles and accompanied by an impressive display of jeeps, tanks, transport trucks, and helicopters. Was this some sort of recruitment campaign gone mad? We never found out.

I had a rare moment of navigational frustration after leaving the Hofburg through an eastern portal. Lisa and I stood on a street corner, each clutching a street map of Vienna. I could find our current location on both maps, although they refused to agree with each other. Stuffing the less detailed of the two maps into a trash can, we walked south along the Augustinerstrasse, where we spied Greene's Titans. In the corner of a small plaza, peering down at us from a roof high above, stood Atlas holding a globe and surrounded by navigational instruments—sextants and telescopes, that sort of thing. In the other corner of the plaza, a stout woman, probably a goddess, also supported a globe. She was surrounded by the sorts of mathematical instruments that we were told to purchase for sixth-grade math class and then never used.

Just around the corner we found the Hotel Sacher. In *The Third Man*, after too many drinks, Colonel Calloway's driver takes Martins to Vienna's most famous hotel, the Sacher, where Calloway's name carries enough clout to get Martins a room for the night. Lisa suggested that we stroll into the hotel to have a little look around, but I was put off by the burly men with grim faces guarding the doors on both sides.

The Sacher was receiving a pretty major face-lift, even as patrons arrived and departed. Distinguished-looking older gentlemen, no doubt very rich, were accompanied by ladies much younger, whom I suspected were growing richer by the hour. The hotel may not be the city's center of indiscretion today, but there was a time when the noble and rich saw it as Vienna's prime destination for extramarital coitus. As we were there in the low season, we might have secured a double room for just €358. If we wanted to splurge, we might have found ourselves in a presidential suite for €3,444. If that all seems a bit pricey, you will be pleased to know that children under six years of age stay free in their parents' room.

We traveled north aiming for the Josefstadt Theater, which plays a pivotal role in *The Third Man*. Here Martins first meets Anna Schmidt, a second-rate Hungarian actress and Lime's girlfriend. Founded 126 years before Greene's time and 216 years before Lisa and I arrived, the theater has hosted innumerable ballet, opera, and theatrical performances. Beethoven wrote a piece, *The Consecration of the House*, specifically for this theater, and, although I have not heard it, I am sure it has lots and lots of notes.

THE NEXT DAY was our chance to see even more of Graham Greene's Vienna. Setting off along Mariahilferstrasse, we found it much quieter on Saturday than on Friday, but the early-morning smell of wet socks was no less powerful than it had been the day before. We wandered down Kärntner Strasse as Rollo Martins had half a century before. Writing in 1950, Greene described the Kärntner Strasse as a ruined shopping district, demolished by wartime bombing, and repaired only to eye level. Well, forget about all of that. Kärntner Strasse is once again a bustling retail mecca featuring super-high-end stuff. If you can find it on Kärntner Strasse, you can't afford it.

Near the core of ancient Vienna, Greene described the enormous wounded spire of St. Stephankirche, towering above the inner city. The church's wounds have since healed. Today the great Gothic spire is one of the great Viennese landmarks, and its tiled roof must make it one of the most-photographed sites in Austria.

While trying to sort out the comings and goings of the final days of Harry Lime's life, Rollo Martins had wandered in thoughtful contemplation along the banks of the Donaukanal. Lisa and I found the canal and crossed at the Marienbrücke. I had always heard that the Danube River was blue, and it may be, but I can say with little fear of contradiction that the Danube Canal is almost exactly the color of French Canadian split pea soup. "Danube Canal Green" is sure to become the latest trendy color in the Martha Stewart line of designer interior paints.

Our destination was Vienna's Prader district, and the Wiener Riesenrad was our big target. Constructed in 1896, the giant Ferris wheel has survived the ravages of time rather well, and is quite a deal at just €7.50 a turn. There was no line-up, and we were admitted to car number eight along with seven other passengers. At one time there were thirty cars on the great wheel, but only half that number were restored after wartime bombing and the resulting fire.

In *The Third Man*, the reprobate Lime agreed to meet Martins at the Riesenrad, and tipped the woman in charge so that they could have the car to themselves. Martins considered pushing Lime out of the car, imagining him as "a piece of carrion dropping among the flies." Even though we didn't have the car to ourselves, Lisa and I celebrated life by smooching at the apex. The wheel took about ten minutes to complete the circuit, and I would have been pleased to ride for most of the remainder of the day. Mainly because of the company. And the smooching.

At one time, the district between the Danube River and its canal was an imperial hunting ground. The Volksprater Funfair arose in the nineteenth century, and today the Ferris wheel is only one of many attractions on offer. We found an endless array of rides and games of chance. Curiously, despite its approaching midday on a Saturday, we found the park almost deserted. The Dizzy Mouse, a small roller coaster, had no patrons, and its larger cousin, the Super 8er Bahn, had only two riders. The Grand Autobahn bumper car ride had only one occupied minicar. This didn't provide a lot of bumping opportunities for the father-son team in it. The Donau Jump log ride, the Boomerang roller coaster, the bungee jump, and the reverse bungee jump were all deserted. The miniature railway stood mute. The haunted house claimed to feature a Sensationen Thriller Tunnel but could not claim any patrons. Lisa asked for permission, and then patted horses at the pony rides. The ponies had been looking really bored before Lisa arrived, as had the ride operator. Most of the food kiosks were closed, and the rest had the faint aroma of food that has been sitting too long. The total scene was pretty creepy. I almost expected an evil clown to jump out at us from behind the Geister Schloss, which looked remarkably like a Ghost Castle.

ARRIVING AT VIENNA'S main train station at a civilized hour on a Sunday morning, we had managed to secure reservations in the nonsmoking section. The seats faced backward, but I have never seen that as a big drawback. Having been a breech birth, coming into the world bum first, this form of orientation has never seemed too far out of the ordinary.

And so, back end first, we chugged out of Austria and into the Czech Republic. At the border, a gang of burly armed Austrian border agents got on the train and stamped our passports to show that we had left their country. A few minutes later a single Czech border agent, not at all burly and with no gun in sight, came by to validate the last page of our passports with a barely used green border-crossing stamp. No muss, no fuss, no fee. "Welcome to the Czech Republic."

The train took us on through agricultural land, up to highlands, and then back down to agrarian flats. Everywhere the trees wore their autumn best and the golden leaves were pure luminosity. The landscape was dotted with woodlots and towns, some with sprawling factories and some without. Despite its recent entry into the European Club of All Things Wonderful, the Czech Republic seemed to have a bit of catching up to do. Other than some recent-model cars, the scene might have been straight out of 1958. We didn't see many billboards or glitzy storefront signs. Streetlights were distinctly low-tech. There were many rooftop television aerials but few satellite dishes. The skies seemed almost devoid of airplane contrails, and in one field, and elderly man gathered rocks, or perhaps rutabagas, loading them onto a wooden cart drawn by an elderly ox.

At the Holešovice train station in Prague, we were besieged by people trying to convince us to stay in one hotel or another; Prague hotels must have a high vacancy rate in late October. For the rest of our stay, we were set upon by people offering to exchange currency for us, a practice we had been told was illegal and possibly dangerous. Indeed, I came to believe that "Do you want to change money?" must be the English expression spoken most frequently and with greatest fluency in Prague.

Also, we had been warned not to trust Prague taxi drivers; a taxi ride is a good way to be parted from your crowns and hellers. There are official rates for travel within the city of Prague, but drivers use all sorts of schemes for ensuring that a tourist doesn't pay that rate. Instead we took the metro, which was clean, spacious, and fast. The fare of 12 crowns translates to about 60 cents. The ticket was a small bit of artwork, with swirly and sparkly Spirograph designs, well worth 60 cents all by itself.

We had dinner in a restaurant attached to a microbrewery. I started with a dark lager and Lisa had 100 ml of *Klasická Staročeská,* something like sherry and something like mead. I ordered the vegetarian platter and found that it was composed of potatoes, dumplings, carrots, cauliflower, and rice, all boiled and largely unseasoned. Lisa bravely ordered *Svíčková na smetaně knedlíky,* and found it to be more or less dumplings and beef in candlesauce gravy. I cannot imagine how I went 46 years without ever trying my second drink, *Kopřivové.* It was Danube Canal green, tasted vaguely of lawn clippings, and was apparently made of nettles. Absolutely fabulous. The whole meal cost 296 crowns, about $15.

LISA TOOK RESPONSIBILITY for flipping through the guidebook and finding wonderful things to fill our day, which she did brilliantly. We started off with a stroll to the Národní Muzeum, a symbol of Czech national pride. It is also home to the two Labrador Ducks that we were scheduled to see the following day. The front doors are at the top of a magnificent ramp and staircase. The main doors are flanked by statues of History and Natural History, sculpted in 1889 by J. Z. P. Mauder. Give Mauder credit for insight. History is a sad-looking old fellow, propping himself weakly on a tablet. He is missing a toe on his left foot and a finger on his right hand. Some sort of allegory, I suppose. Natural History, on the other hand, is a virile and sexy young chap, planning his next exploit on a globe, with all of his digits intact. I have been describing this sort of scholastic dichotomy to my history professor colleagues for years, and now I have Mauder to back me up. The museum's backside is a little less opulent than the front side, being adorned with graffiti that even English speakers would recognize as rude. A number of people sleeping rough were propped up against the building.

And so, with a full day of adventure ahead of us, we turned our backs on the museum. From our elevated position, staring off into the distance, we got a sense of Prague's smog. If there is a European Union standard on air quality, the Czech Republic must be working off a grandfather-clause exemption. From this vantage, and for the rest of our stay, we saw an endless parade of uniformed police officers on the beat, backed up by scores of roving police cars. Prague has a reputation as one of the safest large cities in Europe.

The museum faces Wenceslas Square, to which every visitor to Prague is drawn. Once a center for horse trading and the focus of several important protests against the former communist regime, it has evolved into a consumer strip, a sign of change in the Czech Republic. Western chain stores have moved in to take advantage of the new commercial climate. Smaller shops offered trinkets to lure tourists.

Beyond lay the Old Town, with the Staroměstské Náměstí, or Old Town Square, at its heart. The square's architecture beggars description, with an amalgam of neo-Renaissance, Romanesque, Gothic, Art Nouveau, and Rococo styles. If you aren't satisfied with the Church of St. Nicholas, then you can have the Church of Our Lady before Týn. We gathered to see the chiming of the richly appointed astronomical clock in the Old Town Hall. As the top of the hour approached, members of a group of English-speaking youth reminded each other, "Watch out for pickpockets." At the stroke of 11:00 we watched Death, a Turk, Vanity, and Greed go through their little gyrations.

Time was passing. Breakfast had been served a long time ago. We aimed for the Karlův Most, the Charles Bridge, reported to be one of Prague's greatest tourist stops. We were certain that the neighborhood around the bridge was likely to be filled with restaurants, and indeed it was, but each place had something wrong with it. Either the venue was full, it had a posted menu that included no meatless meals, or had a menu written entirely in Czech. The closer we got to the bridge, the more costly the food became, until restaurants were asking 400 crowns for a small tossed salad. I was getting lightheaded from lack of food, but not dizzy enough to pay that price. We wandered away from the bridge, down a back street, and down another back street, and found a nice pizzeria with very reasonable prices and good food.

Construction of the Charles Bridge began in 1357, and visitors to Prague have been drawn to it ever since. Indeed, unless you are in Prague to drink yourself to death on cheap beer, you are virtually certain to visit the Charles Bridge. At first, only a simple cross adorned the bridge, but starting in 1683, statues of saints were added. I counted thirty-three saints, but may have missed a few. Concentrating on things spiritual was difficult, because the bridge was occupied by a carnival of musicians, jewelry merchants, quick-sketch artists (seven minutes), portrait artists (twenty minutes), and stalls selling sketches and black-and-white photographs of Prague scenes. Like a magpie, I found it difficult to ignore all of the shiny things on offer.

OUR CAVORTING IN Prague was just a lead-up to the real thing—a date with two Labrador Ducks. Arriving at the National Museum a bit early, we looked across at Wenceslas Square and contemplated the sacrifice of Jan Palach, who set fire to himself in 1969 to protest the Soviet occupation. It would take another twenty years for Czechoslovakia to regain its independence in the "Velvet Revolution." Just three years later the nation was subdivided into the Czech Republic and Slovakia.

At the stroke of 9:25, I couldn't hold myself back any longer, and we stormed the museum's doors, only to find them locked. How odd. We found a sign explaining that, "for technical reasons," the museum was to remain closed until 1 p.m. A gaggle of lawyers had rented the facility for the morning. Undaunted, we tried one door after another, until one opened. Inside, a pleasant guard was more than pleased to take us to the reception desk. The lady at the desk called the extension for my contact, researcher and curator Jiří Mlíkovský, and, finding that he had not yet arrived, invited us to take a seat.

At this time the first of the museum's contradictions made itself apparent. Although the building is a magnificent neo-Renaissance construction, there was only one visitor chair, and it bore the remains of an earlier visitor's chewing gum. We chose to stand. While waiting, we counted the number of different types of marble featured in the grand entrance, spotting ten. Despite this splendor, the elevator was out of commission, and had been for some time. Lisa wandered off and found a statue of a princess, and another of the plowman whom the princess wed. While Lisa wandered, I checked my notes and found that Prague has the world's sixty-second-largest collection of stuffed birds.

In contrast to tiny Anita Gamauf in Vienna, Mlíkovský is a giant, with long hair and an enthusiastic beard. It took me a minute to realize he reminded me of musician David Crosby. When Mlíkovský walked, it appeared that his arms were loosely attached to his shoulders, and he swung them with gusto.

Mlíkovský led us through a side door and we began a long ascent to the museum's floor with the most rarefied air. When we got to the top, we followed Mlíkovský down a long series of halls and ramps with more twists and turns than a Gordian knot. I couldn't have found my way back with a map, a compass, a trail of bread crumbs, and a tracker dog.

Labrador Ducks 29 and 30

Mlíkovský deserves hero status. Besides facilitating our visit, he had delved deep into the archives to see what he could find about the male and female Labrador Ducks the museum was justifiably proud of. He found reference to them in the 1842 collection of Baron Christoph Fellner von Feldegg, who had purchased them for 12 gulden. Regrettably, no information was available about when, where, or by whom the ducks were collected. Feldegg died three years later, and seven years after that the Czech museum's Director, Antonin Frič, purchased a portion of Feldegg's huge collection of 4,500 birds. Strangely, the Labrador Ducks were not specifically mentioned. Even so, by 1854 they were listed as part of the collection of stuffed birds safely housed in the Czech museum. The catalogue system there has jumped around a bit over the years, and my drake and hen are now known by the numbers PMP P6V 4052 and PMP P6V 4053, respectively. And if that wasn't enough, Mlíkovský was able to tell us that the Czech term for Labrador Duck is *Kachna labradorská*, although he favoured *Turpan labradorský*, meaning Labrador Scoter.

Prague has long been proud of its pair of Labrador Ducks. The drake and hen are mounted on a single contoured base, the former placed a little higher than his mate. Sand and seashells had been glued to the base to simulate an algae-covered seashore. They were waiting for us in Mlíkovský's office. As soon as I walked in, I said, "Oh, oh!" Lisa jumped in and said, "They're both males, aren't they?" Well, yes, dear, thank you for stealing my thunder. They are, indeed, both drakes. The bird thought to be a female is almost certainly a very young male. Mottled gray and brown all over, there was no evidence of a black neck ring or stripe on the head, and at first glance it certainly looked like a female. Even so, the specimen has a swoosh on his bill that I have found on almost all male specimens, no matter how young. Mlíkovský was philosophical about the revelation. "Well, it's best to know." Both specimens had been given cherry-red eyes, which seemed a little big for their heads. The taxidermist had been overly enthusiastic with the paint, giving both specimens orange legs and toes.

After my examination was complete, Mlíkovský gave us a tour behind the scenes. The Labrador Ducks normally reside in a locked cabinet with two Great Auks, a couple of Huias, nine Passenger Pigeons, a Carolina Parakeet, and a Dodo skull. I took a photograph of Mlíkovský with the Dodo skull, a priceless artifact. These specimens won't be in their present locality for long. The building was 115 years old, and while the front end was maintained beautifully, the back end hadn't been. After years of insufficient care, the research collection is now due to be moved to the outskirts of Prague, to a facility better suited for preservation.

Just before our day came to an end, I went for a run through an early-evening mist that made the cobbled streets slick. I ran down a street that changed its name repeatedly, then south along the Vltava, crossing at a railway bridge and cutting back across the Jiráskuåv Most. I stopped halfway across the bridge to remind myself that I might never pass this way again. "Breathe it all in while you can, lad." My upper bronchial tree burned a bit from the smog. My legs were gloriously tired from our adventures. The bridge deck trembled a bit as each passing car hit an unfilled pothole. I could hear wailing sirens on both sides of the river. The city was gently lit as Prague's one million citizens settled in for the evening. We were five days short of the Czech Republic's six-month anniversary of entry into the European Union. I think great things are afoot.

Chapter Fifteen

Backpacking Through Gotham City

To my way of thinking, a one-hour airline flight is likely to bring you to a close friend. With a little luck, naughtiness will ensue. Family members or business deals are probably waiting at the end of a two-hour flight. You will be doing well if you get a hug or a handshake. If you are on a flight for four hours, you are likely heading somewhere exotic, hopefully with sandy beaches, palm trees, and a never-closing bar tab. In contrast, a nine-hour flight is simply an insult to the system. Nine hours in the air bring on dehydration, anoxia, disorientation, and nausea, particularly since longer flights are almost guaranteed to end with gut-wrenching turbulence and excessive yaw. And yet nine hours in the air were needed to get Jane and me across an ocean to the eastern United States, and a chance to snap up most of the remaining ducks all in one go.

I trust that you have been paying attention. If so, then you will have noticed that my wife's name is Lisa, not Jane. Lisa was home continuing her life-and-death research into the recovery of heart muscle following a heart attack. I was traveling to the United States with Jane Caldwell, a Scottish cardiac specialist who had never been to the U.S., but was keen for the experience. I had absolutely no intention of having a coronary, but even so, I felt a little more secure about life in the presence of a heart specialist. Everyone on Earth has an image of America. Jane was keen to see how America measured up to all the hype.

Before Jane and I set off from Glasgow, several friends had asked me how the United States had wound up with so many Labrador Ducks. I explained that it came down to a combination of geography and money. Labrador Ducks were silly enough to spend their winters in the waters around New York City, where they made good targets for shotguns. Secondly, Labrador Ducks became rare at the time when American museums found themselves with more money than their European counterparts, and so the Americans snapped up every rarity that became available. When it comes to stuffed birds, England has the single biggest collection in the world, but in second, third, fourth, and fifth places are collections in America.

There was just one small problem with this trip to snap up ducks in the eastern United States. It was not exactly clear how many Labrador Ducks were to be found there. The number that Paul Hahn had cooked up forty-five years earlier, something like twenty-four, didn't exactly tally. I was going to have to play it by ear.

JANE HAD SHOWN no reluctance to share rooms to cut costs, but when she saw our hotel room in Washington, D.C., she must have doubted my taste. The wall had holes where repairs to the electrical system had been carried out, the bathtub plug wouldn't keep water in, and the plug hole didn't drain the tub fast enough to keep a shower from turning into a bath. Long past its prime, the hotel seemed to cater mainly to school groups from Posthole, Nebraska, on a tour of the nation's capital. The bedside clock said it was 8 p.m. Our brains said it was 1 a.m. Our bodies said they hated us. We gave in and went to sleep.

Not willing to jump right into an examination of the Smithsonian's four Labrador Ducks after a long journey from Britain, I had scheduled a day of rest, relaxation, and exploration, and so Jane and I were off to tour some of Washington's great attractions, starting with the Mall. This grand, tree-lined avenue seems to be the

American response to the Imam Square in Isfahan or Tiananmen Square in Beijing—built to impress visitors for centuries to come. The Mall, a mile long, with the U.S. Capitol at one end and the Washington Monument at the other, is lined by components of the Smithsonian Institution. The site of endless protests and celebrations, it is also the spot where Martin Luther King Jr. made his immortal "I have a dream" speech.

Having had a good go with her guidebook of Washington, Jane declared that Capitol Hill was a must-see. We found ourselves at the booth distributing free tickets for the Capitol Hill tour. I gather that on a warm day in summer, you need to line up for a tour ticket six months before you are born. Arriving early in the morning on a cold and windy day in March, we had only thirty minutes to kill until our tour.

At 10:15, Jane and I and thirty-eight other lucky ticket holders found ourselves at the marshaling area for the Capitol tour. We marched halfway up a hill, where we were marshaled again. All of our belongings were x-rayed, and then we were marshaled again. We each were given an extensive list of items prohibited in the galleries of the U.S. Capitol, including electric stun guns, martial arts weapons, guns, fireworks, razors, Mace, letter openers, battery-operated electronic devices, hand lotion, perfume, rodents larger than 8 inches, and rabid livestock. Marshaled again, we were led the remainder of the way up the hill, where we entered the Capitol Building by an entirely unassuming west-side entrance. Upon gaining the building, we were marshaled again. Now thirty minutes into our sixty-minute tour, we had yet to see anything of interest or hear anything worthwhile. The Capitol Hill police could probably teach the American armed forces a thing or two about security, but they could learn a lot from Disney about dealing with groups of visitors.

On the day of our visit, a new statue was being installed. Hence all of the best bits of the tour were not on offer that day. Our tour guide did her best. She gestured in the air and asked us to imagine what the rotunda looked like. Rotund, I would have thought. She was a fount of famous dates and names of architects and vice presidents, which tumbled out at a frightening pace. Between the tour and a pamphlet, I managed to scrape together a few details. The original

Capitol was designed by someone named Thornton. It has housed the U.S. Congress ever since moving south from Philadelphia in 1800. In 1814 a bunch of drunken Canadians visiting Washington burned down the whole shebang, leaving just the naked exterior walls. It took four years of reconstruction before the north and south wings were reopened, and a further seven years for completion of the center building to join them.

With the Capitol tour behind us, we were spoiled for choice. Much of the Mall is occupied by the Smithsonian Institution. A legacy of the illegitimate son of the Duke of Northumberland, the Smithsonian is not just a museum, but a tribute to marital infidelity. It is also an amazing amalgam of eighteen museums and galleries, nine research centers, and the National Zoo, making it the largest museum and research complex in the world. The combined collection includes 143.5 million items, one for every 1.94 people in the U.S. The whole shootin' match was overseen by the Smithsonian's secretary, Lawrence M. Small. By all accounts he had been doing a smashing job, but then he had a little slip-up. According to an article in the *Washington Post,* Small had registered a guilty plea to a misdemeanor violation of the Migratory Bird Treaty Act. It seems that his private collection of Brazilian tribal art contained feathers of protected bird species. Oops.

With all this choice, it was a challenge to know where to start. "Ay, it's all six a half a dozen to me," Jane claimed; "I'm not particularly fussed." And so I put in my bid for the National Air and Space Museum. Like every other little science geek growing up in the 1960s, I was fascinated by everything to do with space, and, dear God in Heaven above, this museum has it all. As a child, I would have sold my soul for a chance to visit it. Even at this point in my life, I would have happily sold Jane's soul for the chance. Luckily for Jane, I wasn't asked to. Like the Capitol Building, the museum was free, and our possessions got another jolly good dose of X-ray radiation. When I held my camera to my ear, I could hear it humming.

The items on display really were top-shelf. We saw the Wright brothers' *Flyer* used at Kitty Hawk to make the first sustained powered flight in 1903. Then it was the *Spirit of St. Louis*, which took Charles Lindbergh 3,610 miles on the first solo flight across the

Atlantic in thirty-three hours and thirty minutes in 1927. Near the entrance to the museum we saw *Friendship 7,* which circled Earth three times in 1962, and the *Gemini 4* spacecraft that allowed Edward Higgins White II to make a twenty-minute space walk in 1965 while his companion, James McDivitt, looked on jealously from inside the craft. We were even permitted to rub a small slice of moon rock retrieved by the crew of *Apollo 17.* Given that the six combined missions that put men on the moon had brought back 838 pounds of rock, when this sliver is worn away by the hands of thousands of visitors, there will be plenty of opportunity to replace it.

On display were a backup Mars Rover, a backup Hubble telescope, and a backup Skylab. The originals were, of course, on Mars, in orbit, and in a billion tiny fragments after a fiery re-entry into Earth's atmosphere in July 1979. But for me, the absolute pinnacle of the museum's displays was the command module of *Apollo 11*, the mission that took Neil Armstrong and Buzz Aldrin to the surface of the moon, along with the third guy, who had to stay in orbit in the command module, and whose name no one ever remembers. Along with a good chunk of humankind, I was glued to the television on July 20, 1969, when Armstrong messed up his immortal one-line speech. He meant to say, "That's one small step for a man, (and) one giant leap for mankind," but he forgot the eighth word, and the omission meant that the sentence didn't make much sense. To give the man his due, the third member of the expedition was Michael Collins.

It was time for our next chunk of the Smithsonian. My camera avoided a fourth X-ray, but got a ruddy good cavity search when we arrived to see some natural history. According to the promotional literature, the museum's floor space is greater than eighteen National Football League fields. By my calculations, that makes it a little smaller than twelve Canadian Football League fields, or a shade over one-fifth the area of Vatican City.

This Natural History Museum is a tribute to what can be accomplished with the will to be great and very deep pockets. Jane and I saw only a small portion of the exhibits, but we were gob-smacked with what we saw. For instance, I wouldn't be at all surprised if I was told that the museum's meteorite exhibition was the best in the world. Big meteorites, little meteorites, a 4.6-billion-year-old meteorite . . .

we even saw a vial of diamond dust isolated from a meteorite. The Smithsonian has managed to gather 20,000 of these space travelers.

A guard in the next room admitted that the Hope diamond was probably the most popular exhibit in the museum. "Yeah, and it's probably worth more than everything else in the building put together," he said. When it was first cut, the Hope diamond weighed a whopping great 112 carats, but it has been hacked at over the years, until today it weighs in at a comparatively slim 45.53 carats. It is 25.6 mm long, 21.8 mm wide, 12 mm deep, and surrounded by tourists. It has a faceted girdle and extra facets on the pavilion. I once dated a woman who could be described in exactly the same way. My date was either pink or taupe, depending on the lighting; the Hope diamond is either blue, deep-blue, dark gray-blue, or violet, depending on who is looking at it. Unlike my date, the gem is semiconductive. Exactly like my date, the Hope diamond phosphoresces red under ultraviolet light. The Hope diamond was donated to the Smithsonian in 1958. My date is, presumably, still skulking through the underbrush.

Jane and I went looking for the Smithsonian's display of stuffed birds, but before finding it, we came to the Hall of Mammals. With an impressive 274 mammals, the display occupies 25,000 square feet, or a bit less than half an NFL football field, or the area covered by 131,000 CD jewel cases. Special emphasis had been given to mammals of the African savannah. Every few minutes, the lights in the hall dimmed, and flashes from hidden lights mimicked lightning. Moments later, hidden speakers issued a deep, chest-drubbing rumble of thunder. Small children ran to grab the trouser legs of their parents. They seemed to be listening to some primitive voice that said "run for cover." I found myself reaching to turn up my collar against the imminent rainfall that never came.

After that, the display of birds came as a huge disappointment. With no signs, we had to ask for directions. We were directed to the basement and two dead-end hallways hidden behind an escalator. Overhead light bulbs were burned out, but in the gloom I spotted impressive cobwebs. The whole exhibit consisted of a few archaic cabinets stuffed with locally collected birds, badly in need of a thorough cleaning. Only one other party had found the birds. A father and his two daughters were told by a passing security guard that the Passenger Pigeon on display had been the last one alive. He explained that the bird, a female, had died in captivity at thirteen years of age, after years of failed attempts to find her a mate. Hang on . . . that doesn't sound right to me.

After the guard left, I gave the family the correct details. The last ever Passenger Pigeon, nicknamed Martha, died in 1914 at the Cincinnati Zoological Gardens at the grand old age of twenty-nine. Efforts to find her a mate had indeed proved futile. As an extraordinarily gregarious bird, attempts to breed a single pair in captivity probably would have been futile anyway. When she passed away, Martha's corpse was donated to the Smithsonian. Despite what the guard claimed, Martha was safely locked away in the research collection, waiting for the day when the museum has room for a proper bird display.

THE NEXT MORNING saw me facing my next great duck adventure. Being a gentleman, and wanting to give Jane every opportunity to wake up slowly, I went for a stroll. Even at 7 a.m., I found an awful lot of armed officers in the area around the White House. An antiwar protestor seemed to have settled in for a long wait in a green-and-white tent across the street. He was probably visible from Bush's bedroom window, and it is a tribute to the American state that he was allowed to continue his nonviolent protest in spite of the embarrassment he must have caused the senior administration. Despite all of the security, a gray squirrel ran back and forth with impunity through the fence and across the White House lawn.

It was soon time for our appointment with Labrador Ducks at the Smithsonian. Arriving at the staff entrance, I signed and printed my name, and was directed to the Visitors' Office, where I signed and printed my name again. Collections Manager James Dean came down to meet us. Dean is a lot of human being but, despite his imposing presence, he let me direct the conversation. Perhaps he is a tad shy and more comfortable with birds than with the peculiar people who study them. He explained that he had been at the Smithsonian for twenty-six years, in a position that was just too good to leave.

Labrador Ducks 31, 32, 33, and 34

The Smithsonian proudly counts four Labrador Ducks among its enormous collection of stuffed birds. Two are study skins, and two are taxidermic mounts. The museum's adult male study skin looks a bit like a torpedo, very efficient for storage purposes, but rather shy on artistic flair. He was accompanied by a zip-lock bag containing roughly twenty of his feathers that had fallen out over the years. Two more fell out while I was examining him. In spots, the feathers that should have been white were a bit grimy-yellow. His legs and toes were honey-brown, and the webs between his toes were dark brown and gray, with small pinholes. The museum's female study skin is a little worse for wear, having been transformed from a taxidermic mount to its present condition. Her speckled gray and brown feathers would have made her the pinnacle of camouflage while sitting on a nest. Her right foot is broken and is missing its hind toe, her tail feathers are worn, her neck is bashed up, and she has only one glass eye. Word on the street is that both of these specimens were shot at Martha's Vineyard by Daniel Webster, Massachusetts senator and presidential wannabe. Webster gave the birds to Audubon, who then gave them to Professor Baird, who saw the specimens safely into the Smithsonian.

There are two good stories associated with these specimens. The first is that Audubon used them as models for his life-size painting of *Pied Ducks*. Given the shape the specimens are in, you have to give Audubon extra points for interpretation. I wondered if the pinholes in the male's webs were the result of Audubon's pinning the feet into just the right pose. The second story involves ornithologists Phil Humphrey and the late Robert Butsch. In the 1950s, these gentlemen borrowed the adult male from the Smithsonian so that they could x-ray it and take it apart to see what they could learn about its muscles and bones. Butsch then reassembled the duck as a study skin, and prepared to ship it back to Herb Friedmann at the Smithsonian. Being something of a joker, Humphrey came up with the idea of gathering up some old bits of duck skin and feathers they had lying around, and shipping those to Friedmann along with profuse apologies, explaining that they hadn't been able to reassemble the duck properly after their examination. Better judgment prevailed, and the real Labrador Duck was sent back.

The museum's other two Labrador Ducks are both drakes. The first is an adult, donated by the American Museum of Natural History in 1872. Nothing is known about where he was shot or by whom. For reasons unknown, his tail is particularly worn and frayed. He is a bit greasy around his brown glass eyes and the base of his bill. His feet and toes had been painted battleship gray, and his left foot is a bit mangled. He stands on a block of white foam, and no matter how I turned him, his big brown eyes always seemed to be staring at me. The Smithsonian's final specimen, a young male, was apparently shot on Long Island, New York, in the fall of 1875, possibly by New York City taxidermist John G. Bell. If the information is correct, it provides the poor young drake the dubious distinction of being the last reliable sighting of a Labrador Duck. Like his adult companion, he sits on a block of white foam and is held to the base by delicate white ribbons. Like his companion, the proximal part of his bill had been done over in thick orange paint. When not being poked at, he has a small plastic bag over his head because his feathers are falling out. An ignoble end of the road for a species.

My peeking and poking took three and a half hours. With no opportunity to sit for any of it, I was a little bagged by the end. Dean and I returned the four specimens to their locked cabinet. The Smithsonian counts something like 600,000 bird specimens in its collection, and I asked Dean if he had a favorite. He didn't have an immediate answer, but then opened a drawer to show me a Hudsonian Godwit, a long-billed, long-legged wading bird. Charles Darwin had collected it on East Falkland island. The look of reverence on Dean's face told me that he had a particular fondness for this specimen.

THE NEXT MORNING, Jane and I relaxed as our Amtrak train zipped north, taking us past branches of Chesapeake Bay. According to John Audubon, this region was as far south as Labrador Ducks ever got. The bay was flat, gray, and protected by a line of deciduous trees on the shore. Ducks floated near the bank, but the train was moving too fast to see what sort they were. Almost certainly not Labrador Ducks.

In the past I have moaned about the inability of Amtrak to get me where I needed to be when promised. Well, so far on this trip, they were doing a perfectly good job, and Jane and I found ourselves in Philadelphia smack on time. Still early morning, we dragged our luggage from the train station to Philly's Parkway Museum District, site of the Academy of Natural Sciences, home to five Labrador Ducks.

Before going on to be nasty about Philadelphia's Academy of Natural Sciences, I feel I should give it a really big buildup. The academy is the oldest natural history institution in the Western Hemisphere, established in 1812 for the "encouragement and cultivation of the sciences and the advancement of useful learning." I would be keen to know what constitutes "useless learning." With much fanfare, the institution threw open its doors to the public in 1828. After outgrowing its housing a couple of times, it settled in to its current location on Benjamin Franklin Parkway in 1876. At one time on the outskirts of Philadelphia, the academy is now in the heart of everything important, roughly halfway between the Philadelphia Museum of Art and City Hall. Its scientific collection contains 25 million items, including the first dinosaur skeleton discovered in North America and the only meteorite ever collected from New Jersey. The Philadelphia Academy: big research projects; big education programs—on the whole, a pretty fine place, wouldn't you think?

But then it all starts to fall apart. The short walk from the train station was pleasant enough, but as we approached Logan Square, we faced the disturbing image of a large community of homeless persons living at the center of a fountain, now dry. The academy itself scared me, reminding me of the sort of Home for Naughty Boys that my parents pointed out when I was young as a way of frightening me into good behavior. If not quite a late-nineteenth-century reform school, then perhaps a penitentiary for wayward Walmart employees.

At the main entrance, Jane and I were directed through an improbable series of hallways to a reception desk near the back. There we were issued with the customary VIP visitor passes and waited for the collections manager, Dr. Nate Rice, to gather us up. He led us through the catacomb of nonpublic areas to the bird collection. Rice unlocked the cabinet of extinct treasures to reveal three of the collection's five Labrador Ducks, explaining that the other two were on public display. "Shall I measure these three first, and get to the two on display later?" I asked. Rice got a really embarrassed look and said, "Uh . . ." "Is there going to be a problem, Nate?" I continued. "Uh . . ." continued Rice. Despite having given him nine months' notice of my arrival, and despite my reminder two months before my arrival, and then again eight days before the big day, Rice still hadn't made arrangements for me to have access to the ducks on public display. I settled in to examine the first three ducks, while Rice left to put in frantic calls to folks in the exhibitions department.

Labrador Ducks 35, 36, and 37

When it comes to Labrador Ducks, the folks at the academy seem to have inherited some rather slipshod bookkeeping. And so, after a lot of digging and prodding, here is what I have managed to figure out about the ducks in Philadelphia. Late in the nineteenth century, the academy had three Labrador Ducks—two immature drakes and a hen. Fast-forward seventy years, and a fourth duck appears on the records, this time an adult male. Today the collection includes a fifth duck, another adult male. All five are taxidermic mounts.

More than a century ago, leading ornithologist Whitmer Stone had a good look at the three ducks in Philadelphia. He concluded that the female and the immature male with slightly more white on his breast (catalogue number 5579) were probably from the same collection, as they were mounted in the same fashion. Based on his knowledge of labels and stands, Stone speculated that they were from a collection in the Pennsylvania or New Jersey area, "most likely by Krider or Cassin," who were probably important people in their day. The female was one of the two specimens locked away in the public display area.

The tag around the leg of the immature male in this pair originally had a ♀ symbol, but this had been imperfectly erased and replaced by a ♂ symbol. In ink, the tag also read NO DATA. He is in rather rough shape, with broken tail feathers and a damaged right leg. The glass eyes were in need of cleaning. A repair job on the bill suggests that the taxidermist might have started with an imperfect specimen. Some sort of compound had been applied, presumably to patch up holes, but that material had blistered, leaving the bill warty. A few feathers on the back of his head stuck out, as though the bird was facing into a very strong headwind.

Stone wrote that the second immature male specimen "was procured by Dr. Thomas Wilson, through Verreaux, and was probably included in the collection of the Duc de Rivoli. This bird was presented to the Academy by Dr. Wilson with the rest of his collection." This specimen is known today by catalogue number 5577. Stone wrote about a small label attached to the specimen's leg that didn't have any worthwhile data. I found a small card on the specimen's base indicating that the specimen may have come from the Sanford collection, possibly with a link to Berkley through Beck. All of that may mean something—or nothing. He is, however, the best specimen of the three in the research collection, mounted in a posture that suggests alertness, but with a sense of economy of space. Strangely, his legs and bill are varnished and look almost like carved wood.

The third duck in the research collection, catalogue number 30245, is one of the two adult males. A tag on his right leg suggests that he was acquired from the Carpenter Collection. A similar notation appears in Hahn's 1963 summary. All of this suggests that the adult male on public display is the one the academy picked up most recently, from person or persons unknown. The drake in the research collection had beaten-up feathers and his legs and bill were inexpertly painted and varnished. The taxidermist had mounted his body too close to his feet, and his head too far from his body. The specimen's eye sockets and the base of his bill are greasy, and he was given yellow glass eyes too big for his head, making it appear as though he has an overactive thyroid gland. Poor sod.

I was feeling more and more down as Rice made repeated trips to my workspace without any good news about the two specimens locked up in the public galleries. Rice probably comes from good Christian stock, and may even have done some good and noble deeds in his day, but on that particular day I was just about ready to make disparaging comments about his parentage and his chances of achieving everlasting salvation. It was then that I realized I was coming down with a stinking great head cold, probably something that I picked up on the flight over the Atlantic. With ten days of touring ahead of me, the timing wasn't good.

Labrador Ducks 38 and 39

At this point, there seemed to be nothing else for it. On the third floor of the museum's public galleries, at the end of the endangered species hall, I looked into the glass-fronted display cabinet that contained a handful of extinct birds. It wasn't a big cabinet, depicting a rocky cliff face, with a two-masted sailing ship in the distance. Small it might be, but it contained some great treasures—a Great Auk, two Eskimo Curlews, and my two Labrador Ducks. I felt sick, and it wasn't just because of my building cold. My quest to examine and measure each and every stuffed Labrador Duck in the world looked to be at an end. After thirty-seven Labrador Ducks, here were two that I couldn't examine at close quarters. They were just a few inches away from me, but on the wrong side of a sheet of glass. I could even see the door at the back of the display through which I could gain access, if someone would just hand me the key. I made notes about what I could see from the wrong side of the glass. Their bills appeared to have been painted in black, custard yellow, and baby blue, and the legs were painted gray. The female was better lit than the male, and her tail and flight feathers were less worn than in most specimens.

Then I backed up a couple of yards, sat on my camera case, and watched to see if anyone looked at my ducks. Most of the museum's patrons were members of school groups. Almost all strolled by the display without a second glance. Two skinny boys in baseball caps, a year or two from the start of their college days, looked at the display for a full ten seconds while being teased by a tangle of hormone-fueled girls of the same age. Then a couple in their seventies came by. They spotted the ducks and took a good long look. Perhaps they had the patience of age. Perhaps they were a bit more sympathetic to the plight of Labrador Ducks, being a little closer to extinction themselves. After twelve seconds, they wandered away.

When I left the academy, Rice had my business card in hand, with the telephone number of our hotel and our room number. Rice had promised he was going to try really, really hard to get me access to the display ducks before Jane and I left Philadelphia the following morning. Well, I suppose we all live in hope.

I met up with Jane and we dragged our bags to our hotel, which, luckily, had a bar, where we downed beer and sandwiches for lunch. Maybe I could poison my cold virus with alcohol. Maybe I could drown my unflattering thoughts about the academy. I flirted briefly with the lady behind the bar, most of our conversation circling around professional wrestling.

Looking to improve our image of Philadelphia, Jane and I set off for the Rodin museum. After Paris, if you want to see sculptures by Rodin, Philadelphia really is the place. A wealthy patron, Jules Mastbaum, had paid for the recasting of Rodin's greatest works and the construction of a building to house them, but, regrettably, Mastbaum checked out before the work was completed. We rented headsets to get the best possible Rodin experience. We learned that because of poor eyesight, Rodin had developed into a rather tactile fellow. We saw *The Thinker, The Burghers of Calais*, and *The Gates of Hell*, which I had always suspected were somewhere in Pennsylvania. There were also a couple of fine statues of what can only be described as soft-core lesbian porn. So, overall, a really great place.

My cold was rapidly getting worse, and it seemed as though I had two options. I could go for a run and risk making the illness much worse, or I could pamper myself a bit, catch a nap, and miss the opportunity to see more of Philadelphia on foot. I went running. Being fond of running over bridges, I ran through Philadelphia's central core, aiming for the Benjamin Franklin Bridge over the Delaware River. As I approached, a sign warned cyclists and pedestrians to use the bridge at their own risk. Only sensible, I would have thought. However, the city of Philadelphia had ensured that I would be at no risk whatsoever, blocking all pedestrian access to the bridge with a massive chain-link fence and padlock. Back at the hotel, the little red "message waiting" button on the bedside telephone remained resolutely unlit. Come on, Nate. Show me what you can do!

After a shower, it was time for Jane and me to drown my sorrows and germs further. We found a nice little pub with good beer on tap. We chatted, drank, and watched a couple at the next table go through a dance of seduction. I told Jane that, as a behavioral ecologist, I liked to interpret the behavior of people courting. The young lady was pretty good at it, playing with her hair, and touching his wrist when her companion made a joke. He was pretty hopeless at it, failing to be amused by her little quips, and moving away every time she tried to move closer. By the time we left the bar, his chances for a little action were rapidly approaching nil. Jane suggested that he might be gay; I suggested that he might be dim-witted.

Back at the hotel, I checked the little red light on the telephone one more time. No luck. Thinking that Rice might have left a message with the front desk rather than on the answering machine, I went to ask the fellow at reception. "Is the red light on your telephone lit up?" he asked. I had to admit that it wasn't. "Well then buddy, you haven't got a message."

Jane and I had Saturday morning to discover Philadelphia. In full tourist mode, we sought out the historical quarter of Philadelphia. At Christ Church burial ground, established in 1719, we found Benjamin Franklin's grave, inexplicably covered with coins. The headstone for Gerald J. Connelly Jr. (1927–1991) celebrated his life as a seaman, soldier, and safecracker. Huh? Most headstones marked the remains of persons who had passed away long before Connelly began his life of crime. Benjamin Rush (1745–1813), a signatory to the Declaration of Independence, was described as a heroic physician, teacher, and humanitarian. Captain John Shaw (1772–1823) was recognized as showing "integrity above suspicion, and honor without blemish." Colonel Benjamin Flower (1748–1781) was apparently "punctual." When my time comes, I just hope that my mourners can think of something more complimentary to say than that I managed to show up on time.

To see the Liberty Bell and Independence Hall, Jane and I stood in line for thirty minutes for an overly enthusiastic grope by security personnel. Entirely dissatisfied with their own lives, members of the security group took delight in making senior citizens walk back and forth through a metal detector, removing belts and wrist watches and eyeglasses until they didn't make the machine beep anymore. "But I've got a metal hip," explained one silver-haired lady. The Liberty Bell, placed in the Independence Hall tower in 1753, called members of the Pennsylvania Assembly to debate and vote. Somewhere along the line it developed a God-almighty crack. The bell is now housed in its own center, surrounded by security personnel. An old fellow in a white baseball cap neatly summarized the Liberty Bell for me. "Well, there she be!" It's a big bell with a big crack, and no one gets to ring it anymore.

We moved on to Liberty Hall. "The Birthplace of the Nation," it is the spot where delegates from the American colonies formed the First Continental Congress. As the tour began, a ranger of the National Parks Service told us, "No food, no drinks, no chewing gum—there is a garbage bin to your right. Use it!" He went on in this friendly tone. "Cell phones must be turned off. Stay with your tour group. Take a seat against the wall. I will be checking tickets for the eleven-forty tour!" Jane and I were part of a group of sixty-four. Some of the visitors were from Russia and given explanatory pamphlets. Presumably in Russian. I wanted to ask for a pamphlet in Scottish for Jane, but I was a bit frightened of the pretour guard, er, guide.

This fellow was replaced by a ranger who was to guide us around the site. Undoubtedly, she had auditioned for every comedy club along the Eastern Seaboard, and, failing to get even a single gig, settled for giving talks for the Parks Service. She was abrasive. Had she just stumbled in from killing a bear and not yet found her shampoo or a hairbrush? For a comedian-wannabe, she was surprisingly self-conscious, unable to look any of us in the eye.

The tour began. "You can learn a lot about life from George Washington," she said. I pondered what those lessons might be. Visit your dentist regularly to avoid a mouth full of badly fitting false teeth; or, don't claim to have thrown a silver dollar across the Potomac River when everyone knows the river to be a mile wide; or, don't let your physicians remove five pints of your blood in an attempt to heal you? This ranger clearly felt that being an enthusiastic supporter of America meant being an enthusiastic critic of Great Britain. "I don't want to pick on the monarchy, but let's talk. The Queen knows she's gonna die. Right? So what does she do? 'So long?' 'Good luck?' No! She gives the crown to her son. I mean, let's get real. So, anyway . . ." She didn't seem too keen on the French, either: "So we owe France twelve million dollars for the American War of Independence, and George Washington says, 'So long, guys!' So you know what Washington tells them? I mean, this is true. He tells them, 'No way!' I mean it!" To give her credit, she taught me that George Washington was one of only six persons who signed both the Declaration of Independence and the American Constitution, eleven years later.

Rice never got back to me, and the two specimens on public display in Philadelphia remained unmeasured. I consoled myself with the knowledge that the world's largest stash of Labrador Ducks awaited me just up the coast.

I HAD FACED considerable difficulty booking us into a suitable hotel in New York City. I had used the internet to look for something close to the natural history museum and other touristy attractions without breaking the bank. It seems that "close" and "cheap" are mutually exclusive options in NYC. We were faced with three alternatives. We could book into a youth hostel at seven persons to a room, a flea-bag hotel with fewer stars than Guantánamo Bay, or something way beyond our budget. I swallowed hard and booked us into the final, rather posh, option. Even so, I had to book a room with just one bed to keep costs down. Jane had been pretty good when I told her about sharing a bed.

As we checked in, a fellow in a tuxedo was playing a piano in the lobby. The massive floral display behind the reception desk was real, as were the jade plants in the lobby windows. As we got off the elevator and aimed for our room, a lady in a fur coat pointed at us and said to her companion, "Oh, look, Dolores! Backpackers!" I so badly wanted to say, "Oh, look, Dr. Caldwell! Boors!" I should have—but I didn't.

My head cold was settling in for a long stay, and I was crashing quickly. Jane convinced me to join her in downing a couple of beers in a bar just around the corner from the hotel, which proved to be blessedly free of smokers. The venue was full of happy noise, and to be heard with my croaky voice, I had to stick my lips inside Jane's left ear hole. The beer was good, and probably killed a few germs.

As much as I wanted to discover the Big Apple, I was rapidly running out of steam, while Jane was just getting revved up for a big Saturday night. So while Jane got glammed up for a night of adventure, I crawled into bed. I was awakened at 11 p.m. by a telephone call from a man who insisted on speaking to his girlfriend, Caroline. I explained that Caroline wasn't in my room. "Well, this is room 704, isn't it?" I explained that he had the room right but the occupants wrong. When he demanded to know which room Caroline was in, he heard some very, very rude words, followed by the slam of the telephone handset. I was awakened again at 1 a.m. when Jane couldn't fit her key in the lock.

Over breakfast the next morning, I got the synopsis of Jane's Saturday-night adventures, which began with the words: "I love this city." She had started off with a taxi ride to a bar-and-grill combo. She plunked herself down next to a fellow from Ireland who claimed to be a technical advisor to the Secretary-General of the United Nations and owner of a Paris newspaper. The fellow bought Jane drinks and then a meal. Then the barmaid bought her a drink. Jane was one popular Scot. As the evening wore on, she met a composer, an accountant, the owner of a bowling alley, a pediatrician who also played French horn, and a tall, blond, gorgeous massage therapist. The party moved on to a nearby nightclub. Jane danced to salsa music, which isn't often heard in Scottish pubs.

WHILE WAITING FOR my Labrador Duck encounter on Monday morning, Jane and I made our Sunday turgid. We started by walking east to Central Park, clearly one of North America's greatest city parks. I had expected the park to be flat, like much of Manhattan, but it has been sculpted into beautiful undulations. Even early in the morning, Central Park was full of recreational runners as well as the participants in the Colon Cancer Challenge Race. We stumbled across a rainbow-clad group of cyclists staging for a ride. I asked the prettiest one (well, why not?) how far they were going. "New Jersey," she explained. You have to be impressed.

The promotional magazine in our hotel room listed some of the fun activities available in Gotham for the out-of-towner, including the opportunity to see a taping of *The Late Show with David Letterman*, and it sounded like a fun thing to do. Jane and I walked south from Central Park to the Ed Sullivan Theater, which looks a lot more impressive on television than on the street. We filled out forms and had a briefing with assistant Seth, who gave us a 50–50 chance of being called for the show the following night. Regrettably, the red light on our hotel room telephone stayed as resolutely dim as the one on the hotel telephone in Philadelphia. Jane and I would have to make our own fun.

We recharged our batteries with coffee and headed for Times Square. What a sensory onslaught! A concrete canyon with bright lights and flashing signs. An army of perceptual psychologists had been working around the clock to figure out how to assault visitors into buying stuff or going places or doing things, none of which they really needed or wanted.

It was roughly at this point that I lost Jane. I looked in the immediate vicinity and then further afield, peering over and around people as best I could. No Jane. Before leaving Scotland, I had promised Jane's mother that the one thing I certainly would not do was lose her daughter. A couple of minutes passed. Still no Jane. My mind started to race. Whom did I contact first—the British Consulate or the Canadian Consulate—and how long did I have to wait before I started panicking? Not much longer. I once read that twelve people are murdered in New York City each day. Surely a good portion of those are Scottish cardiologists. Would the NYPD think that I had killed her? Would I spend the rest of my miserable life rotting in an American prison? Oh, dear God.

"Hi, Glen." She had wandered off to take a photograph.

Feeling the need for a sense of perspective, we got in line for the greatest view of NYC—the observation deck on the eighty-sixth floor of the Empire State Building. It was early on a chilly Sunday morning in March. Surely if there were any line up at all, it would be a short one. Here is my advice—unless you really, really want to see New York from 1,000 feet up, spend the $13 on a round of coffees for your friends. Folks will be in line for the Empire State Building long after you have finished your drinks. But of course no one was polite enough to tell us that. We started off with an escalator ride from the ground floor to the basement, generously described as "the concourse." The building's management had clearly stumbled on a very good deal on ugly yellow paint. Unlike the building's elegant lobby, rich in marble, the basement had been decorated in Early Demolition. We cowered in a long line beneath ductwork, naked light bulbs, and bare-ended wiring. There were missing acoustic tiles, and, although I'm not certain, I think I saw asbestos insulation. All of this was just while lining up to buy tickets, and it snaked back and forth like a bank queue on heroin. Hawkers tried to sell us NY City Passes, NY Skyrides, and Tony's New York Stories, in all possible combinations and permutations. My camera, still glowing from our time in Washington, was x-rayed again.

Having purchased our tickets, we got in another snaking line in a basement hallway with poor lighting. After thirty-five minutes, we took an escalator past the elegantly appointed ground floor to another line in another grotty hallway. This was the queue to have our tickets taken, which led to another line to have our gear x-rayed again. We were finally allowed to board an elevator to the eightieth floor, where we lined up to be photographed in front of a picture of the Empire State Building. This left us in one more line, this time for the elevator to the observation deck. An assortment of audio tours were hawked again. "Last chance!" A hot dog stand or a coffee kiosk would have gone over much better at this point.

We thrust ourselves back out into the daylight, eighty-six floors up, to confront a view like no other: the ultimate array of endless architectural erections, home to millions of people who lived, played, loved, and died below us. I looked deep inside myself, and then out over the city, for inspiration, wondering what would make people pay $13 and stand in line for so long to see it. If it were an eighty-six-story view down on the grand creation of virgin Brazilian rain forest, I'm not sure that it would have been as popular as the Empire State Building. Could it be that this was one of the world's best views of what humankind is driven to create by urges that we cannot fully understand?

Being a little less philosophical, and a little more task-oriented, it occurred to me that a good chunk of all the Labrador Ducks on my quest had been shot within the panorama before me. Indeed, the very reason that Labrador Ducks went extinct may have been staring at me. It seems completely unlikely that my ducks had become extinct because they were harvested at an unsustainable rate. It seems a lot more likely that their numbers had spiraled down because of the increasing number of inhabitants of the eastern American seaboard in the 1800s. All of those people, making all of that sewage, all of it going untreated into the ocean, exactly where the Labrador Duck was spending the winter. For millennia, they had passed the nonbreeding season feeding on mussels in the shallow waters just off the American East Coast, but then the human population exploded. I contend that my ducks were polluted into oblivion.

Back at street level Jane and I separated. She was interested in opportunities in the immediate vicinity, including art galleries and the like. I had an appointment with an intersection nearer the south end of Manhattan. The sun did its best to shine on me by peeping between skyscrapers along the Avenue of the Americas. En route, I met Shopping Cart Man, who stood in the middle of the avenue shouting, "Either you're a Republican, and you're *for* America, or you're a Democrat, and you're *against* America. You can't go on dreaming anymore, Democrats!" Other than drivers who swerved to narrowly avoid hitting Shopping Cart Man, I was the only one to pay him any attention at all. Approaching Washington Park, I spotted Yoko Ono, wife of the late John Lennon. She was reading the apartment vacancy ads in a free community newspaper while waiting for a bus. She was speaking Spanish. To herself. At least I think it was Yoko Ono.

In the 1800s, Mr. John G. Bell operated a taxidermy shop at the intersection of Broadway and Worth. Quite the destination in its day, Bell counted John James Audubon and Teddy Roosevelt among his patrons. Not surprisingly, given the location and Bell's profession, several Labrador Ducks are known to have passed through his hands, including the beautiful drake at the Royal Ontario Museum in Toronto and several fine examples that I was scheduled to see the next day.

Broadway and Worth may have been an interesting place in its day, but nowadays it ranks as one of the three ugliest intersections in North America. On the southwest corner stands the Steps Clothing Company (EVERY ITEM $10), and a Japanese-Chinese restaurant. On the northwest corner is a Strawberry Clothing store and a B'way Best Gourmet Farm. It isn't really a farm. On the northeast corner are a thirteen-floor apartment block and the Independence Community Bank, featuring gargoyles with bat wings. None of these buildings has any apparent redeeming features, but each is a gothic cathedral compared to what stands in the southeast corner. The Jacob K. Javits Federal Building occupies an entire city block, and does so without grace. Forty-four stories by my count, it did as little to satisfy my soul as alcohol-free beer. This is one of the ugliest buildings in the world, so I just had to take a photograph.

This was a mistake. I walked closer and closer to the building across its courtyard, which was deserted on this Sunday afternoon, trying to get a camera angle looking almost straight up that would show most of the building's magnificent homeliness. And then, just before clicking the shutter release, I heard shouting. Although at a distance, this was shouting at its best. I looked around and spotted a security guard, wildly gesturing and shouting. Since there was no one else in view, I assumed that he was gesturing and shouting at me. He motioned me over to his kiosk; if I was going to cause trouble, he wanted to be close to a telephone.

Up close, his dull face was in a state of extreme agitation; clearly he wanted me to speak first. "Good afternoon," I said using a British-tinged Canadian accent. "Can I help you?" It sounded like a pretty disarming thing to say. He explained that the Jacob K. Javits Federal Building was a Federal Building, and that taking photographs of a Federal Building was against the law, and therefore I was in Big, BIG Trouble. "So what do you think you're doin'?" he asked. Well, at that point I was thinking fast. Recognizing a dim-witted bully when I saw one, I also recognized that in an era of American paranoia, I might be in some actual trouble. At the very least, I might have my camera confiscated, and with a couple of dozen photographs of Philadelphia ducks on the film inside, I didn't want to lose it.

And so I decided to tell this fellow my story in mind-numbing detail. He heard about Labrador Ducks, about John G. Bell, about nineteenth-century taxidermy, American natural history museums, about the talks that I give to interested groups, and about how these talks were illustrated by photographs I took along the way. He also heard that I hadn't actually taken a photograph of the Jacob K. Javits Federal Building; he had interrupted me just before I had tripped the shutter.

"I saw you takin' pictures. I saw you. You are in Big, Big Trouble. Big, BIG Trouble." So I played the trump card he hadn't expected. I pulled out my Canadian passport, and said, "Here. Look at this." I suspected that if he had taken my passport, or written down any details about who I was or what I was doing, he would be forced to spend the rest of his Sunday afternoon filling out an incident report. It worked. “I don’ wanna see that! I don’ wanna see that!” he said. After telling me twice more what Big, BIG Trouble I was in, he sent me on my way. Later investigation on the Internet made it abundantly clear that taking a photograph of an American federal building is/is not against the law. Take your pick.

IF THERE WAS to be one really big day in my Labrador Duck quest, this would be it. After dashing hither and yon for years, measuring one specimen in Belgium, a pair in Austria, and a pair in the Netherlands, I was now poised to tick off a stunning eight Labrador Ducks in one building. This would be three birthdays and two Christmases rolled into one.

Surely there is no one in New York who doesn’t know where the American Museum of Natural History is situated, even if they have never been inside. With a postal address like Central Park West at Seventy-ninth Street, the city block on the Upper West Side occupied by the museum must be among the most valuable chunks of real estate in the world. Unlike the foreboding Academy in Philadelphia, the monolithic AMNH is set back from the street and built of a light stone that fairly screams, “Come in! Fun stuff is happening inside these walls!”

Shannon Kennedy, a scientific assistant in the Department of Ornithology, led us to the deepest inner sanctum of the museum’s research collection, where its most precious specimens are kept. Shortly afterward we were joined by Collections Manager Paul Sweet. He assured me that *all* of the institution’s Labrador Ducks would be made available to me, even the four on public display. He seemed like the sort of fellow you would want backing you up if you were trying to capture a rattlesnake. Sweet was endlessly polite to Jane and me, making him a good guy in my books. As I settled in to work on the Labrador Duck study skins, Jane disappeared back into the maw of New York City.

Labrador Ducks 40, 41, 42, and 43

I poked and prodded and snapped photos. If the AMNH had just these four ducks, it would easily be the best collection in the world. They were simple, mute representatives of their kind. A study skin, specimen 45802 shows the combination of gray breast and belly and white throat and upper neck that mark the immature male. He is unique in having tags attached to both feet, and a third tag attached to his bill by a bit of string. It is a beautiful representation of the preparator's art. Almost as beautiful is specimen 45803, an adult male study skin. I found him a little greasy around his eyes and on his right wing, but otherwise perfect, with immaculate white feathers. The second adult male study skin, specimen 734023, was stuffed a little full for my liking, particularly on the forehead, making it seem that he was suffering from some disfiguring brain condition. The specimen was heavier that I would have expected, and I wondered what he might have been stuffed with, if not the usual cotton batting. The final study skin, number 45802, has been described in most publications as a female, although the white feathers of the throat and the uniform gray belly tell me that it is very likely an immature male. The region around his cloaca is a bit mucky, and his bill has a couple of small holes, attributable to the shotgun pellets that brought him to an end. Otherwise, like the three specimens before him, he is spectacular. I was grateful that no one had given into temptation and obscured details of the bills and feet by painting them.

When I finished, Sweet was still waiting to hear from a representative of the exhibitions staff who could open up the Labrador Duck display, so I headed off to the library. I wasn't worried. I kept telling myself that I wasn't worried. Sweet knew what he was doing, right?

Wading through the museum's 1889–1890 annual report, I found reference to a promise of construction of the Labrador Duck diorama by Jenness Richardson of the Taxidermic Department. In the 1890–1891 report, Richardson proudly reported that the Labrador Duck display had been completed. When he bragged once again about completion of the Labrador Duck display in the 1891–1892 report, I began to suspect that the taxidermic department didn't have enough to brag about. The 1890–1891 report also contained an interesting description of work that had been going on in the mammal exhibition. It seems that a great Indian rhinoceros named Bombay had died, and the Taxidermic Department had set about to stuff him. The first step was to remove his skin, which weighed 750 pounds. The skin was then placed in an antiseptic solution for four years to preserve it. Two men then worked for two months to scrape away at the inside of the skin to reduce it to a thickness of a fifth of an inch. This was then mounted on a wooden framework. So, no matter how bad your job gets, remember that your boss hasn't asked you to spend two months scraping the inside of a rhinoceros hide.

Labrador Ducks 44, 45, 46, and 47

At 3 p.m., Sweet came into the library and announced in a loud voice, "We're in!" We dashed to the Chapman Memorial Hall of North American Birds, where a member of the exhibition department had removed the sheet of glass from the front of the Labrador Duck display. The display is an attractive one, meant to represent a winter scene at the margin of Long Island, with one duck swimming and the remainder sitting or standing on the shoreline. The group contained an adult male, two immature males, and one female. Except where a sheet of glass was used to represent water, the base of the display is covered in artificial snow. As I prepared to stick my face and arms in to measure the ducks where they sat, Sweet wondered if it were possible to pull them out of the display where they had resided for the past 115 years. He grabbed the adult male mounted in a swimming position, and POP! Out it came. He handed it to me while he went in search of a cart that could hold all of the ducks. The fellow from the display department went in search of a card to place in the display, explaining that the specimens had been temporarily removed for research purposes. So there I stood, in the public galleries, with a stuffed Labrador Duck in my left hand and a little innocent mischief in my heart.

In 2002, Wayne Gretzky, the greatest hockey player of all time, was general manager of the Canadian men's team at the Winter Olympics in Salt Lake City. Before the games began, Gretzky convinced the arena's ice makers to embed a Canadian one-dollar coin under the spot that marked center ice. The Canadian women went on to win gold. After the men managed the same feat a few days later, Gretzky went to center ice and dug up the coin. The coin now resides in the Hockey Hall of Fame in Toronto.

So, standing in front of the world's greatest public exhibition of Labrador Ducks, I wondered if I could pull off a similar stunt. I reached into my pocket in search of a Canadian one-dollar coin. I didn't have one, but I did find a Canadian dime. It would have to do. Looking around to make sure that no one was watching, I inserted the dime under the artificial snow and smoothed it over, leaving no trace of my naughtiness—except now that I have confessed my actions in a book.

When Sweet returned with the cart, we took the ducks back to the rare birds room, where I made an examination in world-record time. If I had been working at my leisure, the whole thing could easily have taken four hours. I managed to work through them in ninety minutes, so everyone could leave work on time. My task was a bit easier, because the bills and feet of all the ducks in the display had been painted in exactly the same way. It wasn't pretty. The distal half of the bills of each had been painted jet black. A thin line of baby blue paint had been painted along the midline of each bill. The remainder of each had been painted bright orange, except for two of the specimens with lime green paint in their nostrils. I snapped a photo of the group on the cart and, removed from the context of their artistically prepared display, they looked to be sharing a narcotic frenzy.

After years and years of digging through old literature, discarding rumors, and following hunches, I am now in a position to provide you with a brief history of every Labrador Duck that has ever been through the doors of the American Museum of Natural History in New York City.

1. Until 1872, the AMNH had an adult male from the collection of D. G. Elliot. That specimen is now in the Smithsonian in Washington. I had examined that specimen a few days before.
2. From October 17, 1921, to July 27, 1965, the AMNH had an adult male on permanent loan from Vassar College. It

had been in the collection of Jacob P. Giraud, first mounted by taxidermist John G. Bell in 1867, and later remounted by George Nelson in 1921. I had examined that specimen at the Royal Ontario Museum in Toronto some years before.

3. In 1931, D. L. Sandford of the AMNH managed to sweet-talk the Boston Society of Natural History out of an immature male Labrador Duck, which he immediately shipped off to the Forschungsinstitut und Naturmuseum Senckenberg in Frankfurt in exchange for a female Bonin Islands Grosbeak, also extinct. I had examined the duck specimen the year before in Frankfurt. While at the AMNH I also got a look at the grosbeak. I wish that I could say she was pretty; she was dull brown everywhere, with an outsized beak and head that only a mother could love.

Now on to the Labrador Duck specimens still found at the AMNH. The first four are study skins, locked safely away in the museum's vaults.

4. An immature male, catalogue number 45802, was collected off Long Island around 1865, and purchased in NYC's Fulton Market by George N. Lawrence. The AMNH acquired it in 1887.
5. An adult male, catalogue number 45803, also from the collection of Lawrence, was acquired by the AMNH at the same time as the previous duck. It was collected off Long Island ca. 1842 and passed through the hands of taxidermist John G. Bell.
6. Another adult male, catalogue number 734023, shot at La Prairie, Quebec, in the spring of 1862, was purchased by William Dutcher from a Mr. Thompson for $125. Walter Rothschild, whom we have already met, purchased the duck from Dutcher, and then sold it to the AMNH.
7. An adult female, catalogue number 734024, was probably collected from the waters around Long Island. It was sold by John G. Bell to Dr. Henry F. Aten, who then passed it on

to Gordon Plumber of Boston, who sent it to Walter Rothschild, who sold it to the AMNH.

The remaining specimens are taxidermic mounts, which grace the diorama in the museum's Chapman Hall of North American Birds.

8. An adult male, catalogue number 3738, was collected off Long Island around 1862. D. G. Elliot obtained it from Brooklyn taxidermist John Akhurst, and eventually passed it along to the AMNH.
9. Another adult male, catalogue number 3739, came from the collection of German naturalist Prince Maximilian zu Wied-Neuwied. The legs of this specimen were replaced by New York taxidermist John G. Bell.

10 and 11. An adult female, catalogue number 3740, and an immature male, catalogue number 3741, were both collected off Long Island, and passed from John G. Bell to D. G. Elliot, and then on to the AMNH.

12. A female, catalogue number 45801, collected off Long Island around 1842, and sold by John G. Bell to George N. Lawrence, was acquired by the AMNH in 1887.

"Aha!" I hear you say. "Those numbers don't add up." I told you that there were eight Labrador Ducks at the AMNH, but have listed four study skins and five taxidermic mounts. That is because one of the ducks on public display had been stolen. Mary LeCroy of the AMNH filled me in on what was known of the theft. About thirty years ago, outside contractors were in the museum to install pipes in the ceiling in the area behind the bird dioramas. The contents of all of these displays are very difficult to get to—a large, awkward pane of glass must be removed from the front of each. The exception was the Labrador Duck display, which had an access panel in the back. Regrettably, that panel had been left unlocked. It seems that one of the workers, or a visitor to the museum who spotted an unlocked door, spied the access panel, reached in, and strolled off with the duck closest to the back, one of the two adult males mentioned above. It has never been recovered, so the display today contains just four Labrador Ducks instead of five. In Mary's estimation, the thief probably didn't realize the value of what he or she had snatched. That duck may still be on someone's mantelpiece, just waiting to be spotted by a clever bird enthusiast.

JANE AND I had agreed to meet in a bar we had spotted close to the hotel. I was in a fabulous mood, having snagged 15 percent of the world's population of Labrador Ducks in a single day, so I got a head start on Jane, slugging back a Bass Pale Ale. Only then did I realize that I had missed lunch, which probably explained why the sight of the bottom of my beer glass left me feeling loopy.

Jane arrived, and we both tipped back a Groundhog Cider or Woodchuck Cider or something like that. I told her about my success and she filled me in on the impressive array of things she had seen and done that day while I sat at a desk in a windowless room. For Jane it had been a whirlwind of the finest art galleries, headiest coffee shops, and most peculiar museums. We ordered another round of something alcoholic; Jane said that it was Golden Monkey beer, but she may have been teasing. The glass was blurry, but its contents were cold and wet, and so we toasted our success at being wonderful human beings in a wonderful world. At one point, Jane asked what I was thinking, as I appeared to be staring blankly into space. I explained that I was trying to remember enough of my college physics to work out the engineering of our barmaid's brassiere; it really was a stunning piece of work. I must have been getting as blurry as my glass, because Jane politely suggested that food in my stomach would be a good idea.

She firmly took my arm and suggested Vietnamese food. At this point I probably would have agreed to filet of Stonehenge. Claiming that I couldn't read Vietnamese, I had Jane order for both of us. I sorta remember drinking Vietnamese beer, but it may have been Budweiser. As we sat and chatted, we contemplated my latest profound insight into the human condition. It seemed to me, at least at the time, that every adventurous soul should attempt to spend a year living in each of New York City, London, Paris, Mexico City, and Tokyo.

THE NEXT MORNING found me in my running togs in Central Park. At 6:30 the park had more runners than a seniors' complex has liver spots. Almost all of them were running through the park counterclockwise, so I ran clockwise. I wasn't suffering a headache. Honestly.

By lucky coincidence, the New-York Historical Society museum, just next door to the Natural History Museum on Central Park West, was showcasing an exhibition entitled *Audubon's Aviary*. With really no idea what to expect, I was delighted to find three dozen of Audubon's paintings on display. These were not images sliced out of his double elephant folio of North American birds, but the actual originals. To me the best of the lot was the painting of the Carolina Parakeet, not because it represented an extinct bird, but because the portrait had been executed so well. Completed in watercolor, pencil, pastel, ink, and gouache, they served as the basis for the engravings for Audubon's most noted published works. Some paintings had motion detectors nearby, so that an approaching viewer would set off a broadcast of the bird's song or call. Also on display were assorted ephemera, including Audubon's gun, field hat, paint palette, and writing slope. One of the society's volumes of the double elephant folio was open to the page featuring the Great Blue Heron, rendered lifesize. A highlight for me was a painting of Audubon by his son John Woodhouse Audubon. This was the son who had accompanied John Senior to Labrador, and had reportedly found Labrador Duck nests on the hillside that Lisa and I had visited five years before. John Junior must have used some tricky pigments, because as I changed my position, small points of light twinkled off his father's face and hair.

I asked a guard if the society had other Audubon originals. My head swam with thoughts of seeing the great man's original Labrador Duck painting. Her response was, "Yup." When I stood waiting for elaboration, she went on something of a tirade. "Everyone wants to see the Audubons. They don't wanna see nothin' else. We got other stuff, but it's always, 'Where's the Audubons?' They walk in, they look at the Audubons, and then they leave. They were supposed to be done by now, but everyone wants to see them." Not really a big fan of Audubon, I suppose.

As it turns out, the New-York Historical Society does, indeed, have the original Audubon painting of the Labrador Duck. Unfortunately, I couldn't see it on that day, because the curator in charge wasn't on site. Alexandra Mazzitelli, a research assistant at the museum, explained that the painting measures $21\frac{3}{8}$ inches by $29\frac{11}{16}$ inches. The depiction is a combination of watercolor, graphite, pastel, oil, charcoal, and black ink on paper. The painting has "No. 67 Plate 332./Pied Duck" written in pencil in the upper-left corner. Regrettably, the painting had been glued onto a thin cardboard backing after engraving in England and before it was shipped back to the United States. There are faint inscriptions on the backside, but these are visible only as ghost images through the paper, and therefore illegible. Along with the other originals, the Labrador Duck painting was purchased from Audubon's wife in 1863 with funds raised by public subscription. With luck, my next visit to New York will include a stop to see it.

It was abundantly clear that Jane wasn't keen to leave New York. It seemed to be her kind of town, celebrating everything that she held dear. If Jane were to announce that she was moving to New York, I would be surprised not one bit.

I AM A big fan of Mark Twain. I have read all of his major works and quite a few of his lesser pieces. One of Twain's entirely less satisfying later novels is *Tom Sawyer Abroad,* almost certainly written to help keep up with the mortgage payments. In it, Tom, Huckleberry Finn, and ex-slave Jim travel the world in a hot-air balloon. Part way through their journey Tom and Jim have an argument about metaphors. Tom tries to illustrate a point by making reference to the adage "birds of a feather flocks together." Jim, being a little more literal, believes that he has caught Tom out, explaining that bluebirds and blue jays are never found together, despite similarities of plumage. Huck, siding with Jim, knows that the argument has come to an end "because Jim knowed more about birds than both of us put together. You see, he had killed hundreds and hundreds of them, and that's the way to find out about birds. That's the way people does that writes books about birds, and loves them so that they'll go hungry and tired and take any amount of trouble to find a new bird and kill it. Their name is ornithologers, and I could have been an ornithologer myself, because I always loved birds and creatures."

I had always assumed Mark Twain had lived his life in Missouri and Mississippi. However, the people of Elmira, New York, are proud to point out that their fair city played a big role in Twain's life. Given the theme of my journey, it isn't surprising that there is also an important connection between Elmira and Labrador Ducks.

The short version of the story goes that the very last Labrador Duck, an immature male, was shot in the waters around Long Island in 1875; its body resides at the Smithsonian. Less than ninety years after it was first described, that was that for the poor Labrador Duck. More or less. Perhaps. The longer version of the story is more compelling, but considerably less reliable. In 1879, the journal *American Naturalist* published a three-sentence report by W. H. Gregg. Elmira's pharmacist and health officer, Gregg explained that a Labrador Duck had been shot in the town on December 12, 1878, but offered no further details.

Further details were provided in a paper published thirteen years later. According to Dr. Gregg's recollections, the Labrador Duck had been shot by "a lad" when the Chemung River had overflowed its banks as the result of a winter storm. Before Gregg could get to it, the duck had been eaten, and nothing remained but the head and a bit of the neck. Gregg held on to this artifact for years, but it was subsequently lost. A good story, but without a scrap of corroborating evidence.

And it all seems so terribly unlikely. The Labrador Duck was a seabird, and Elmira is more than 185 miles from the ocean. It's more than 60 miles from both Lake Ontario and Lake Erie, the closest bodies of water big enough that a confused duck might have mistaken them for an ocean. Even so, I was spurred on, ready to take Jane on a 400-mile side trip, by the fact that the American Ornithologists' Union recognizes the Elmira record as possibly legitimate, giving 1878 as the end of the road for the Labrador Duck.

Pulling our rented car into Elmira, our first stop was 333 E. Water Street, where Gregg operated his pharmacy. The drugstore wasn't there anymore, of course, having been run over by the thoroughly modern Elmira Savings Bank building. However, just down the street is the Riverwalk Café (Open for Business) and Personnel Images (empty, but Open for Offers to Lease). A stone above the door indicated that the building had been erected in 1842. Not satisfied with the job, a different stone told us that it had been erected again in

1869. Either way, the building had been in place when Gregg had taken time out of his Christmas preparations to write about the Labrador Duck discovery.

It is one thing to visit a town and see where things used to be, but quite another to visit a town and see where things still are. So we crossed a bridge spanning the naughty Chemung River, which flooded so horribly in 1878 and in whose overflow the last ever Labrador Duck had reputedly been shot. The river was very well behaved on the day Jane and I visited, staying well within the enormous concrete walls that had been constructed to contain it.

The good people at the Chemung Valley History Museum were able to fill me in on some important aspects of Mark Twain's life. Not only did Twain spend a good portion of his life in Elmira, he married a local girl, Olivia Langdon. According to the museum, Twain wrote all of his best works in Elmira, including *Tom Sawyer* (1876), *The Prince and the Pauper* (1882), *A Connecticut Yankee in King Arthur's Court* (my favorite, 1889), and *Huckleberry Finn* (1892). Twain shuffled off this mortal coil in 1910 in Reading, Connecticut, but his remains were carted back to Elmira for burial.

The museum houses Twain's chaise longue, his portable writing table, and his wife's wedding gown. It also has the author's typewriter, and somewhere in the deep recesses of my mind I remembered reading that *Tom Sawyer* was the first novel ever to be composed on a typewriter. Then it occurred to me that Twain may have known Gregg. They may have even chatted about the Labrador Duck while Twain waited to have a prescription filled.

Jane set off for an hour's search of fun things to do in Elmira, while I got on with a long-anticipated appointment. A couple of months earlier, I had written to the mayor of Elmira, Stephen Hughes. According to the city's website, Hughes had been born and raised in Elmira. He and his wife, Linda, had been married seventeen years and had three children. Hughes had been elected councilman just seven years after graduating from Southside High School, and served five terms in that capacity before being elected mayor. He apparently enjoys golfing, and can be counted on to show up to see his daughter at cheerleading events. Judging by his photograph, he has a nice haircut and looks good in a dark suit. Just the sort of person who might want to chat about extinct ducks. Hughes had written back, saying that he would be pleased to meet with me.

A sign on the building at 317 E. Church Street indicated that it houses Elmira's police department, its city courts, and the city offices. When I asked for the mayor's office, a police officer gave me a slightly disgusted look, and said, "Upstairs." Odd sort of behavior for a civil servant. On the second floor, I found a uniformed guard and asked where I would find the mayor's office.

"Well, it's upstairs, but I don't think he's there."

"That's fine," I said. "My appointment isn't until two o'clock."

"Yeah? Well, that may be, but I don't think he's coming back. He resigned this morning." I responded with a slack jaw. "Yup," he continued. "About three hours ago."

Good God. How little did this guy want to speak to me that he was willing to quit his job? All right—I'll bite. "Why?" It seems that the mayor had sought to top off his salary with a second job. Unfortunately, this new non-major job would put him into a potential conflict of interest situation over the dispensation of funds. The only sensible option, apparently, was to toss in his job as mayor. And to stand me up.

I walked up to the third floor anyway, if only because I suddenly found myself a bit shell-shocked and without anything else to do for an hour. Coming down the stairs was a friendly, if harried, looking gentleman with his arms full of paperwork. He was sporting a very worn yellow baseball cap and a turquoise sweater with the sleeves pushed up. "Is there anything I can do for you?" he asked. I explained about my appointment with the mayor and my long journey to keep it.

Terry McLaughlin deserves extra credit as a good and decent human being. As Deputy Mayor of the City of Elmira, he had probably been up to his armpits in messy meetings and interviews since sunrise. He could easily have said, "Look, we're having a kind of rough day here, so if you don't mind, I'm going home." He didn't. Instead, he invited me to join him in his office for a chat about Elmira and dead ducks. McLaughlin sat behind his desk and offered me the facing chair. He explained that he had served the people of Elmira for ten years as Councilman and eight years as Deputy Mayor. I got down to asking him the questions that I had prepared for Mayor Hughes.

I got the expected answer when I asked if McLaughlin had heard of the Labrador Duck. "But I do have two Labrador retrievers!" He got half marks for that answer. Asked if hunting was popular in the area, he said that people hunted white-tail deer and black bear, and that fishing was very big, with special emphasis on trout, bass, walleye, and tiger muskies. Terry was particularly fond of hunting in the Adirondacks, where he "shoots black powder." Probably a lot safer than trying to shoot bears.

I asked what the citizens of Elmira were most proud of. He thought for a minute, and offered up, "A small-city atmosphere, with a connection to the past." The streets of Elmira are safe, and the people very neighborly. The winters are generally mild, the surrounding hills beautiful, and there are plenty of recreational opportunities. You can still buy a four-bedroom house in Elmira for under $80,000. Whatever might be said about Elmira, McLaughlin claimed that "no challenge is too large that we cannot overcome it," which seems just about the right attitude for a politician.

McLaughlin gave me an impressively long list of famous Elmira alumni, which included, not surprisingly, Mark Twain, described as "a big tourist gig." I heard about Ernie Davis, who in 1961 was the first black athlete to win the Heisman Trophy. Hal Roach, director of the Laurel and Hardy films, was an Elmiran, as were fashion designer Tommy Hilfiger, and John W. Jones, who was involved in the escape of 860 runaway slaves via the Underground Railroad. In 1999 Elmiran Lieutenant Eileen Collins became the first woman to command a space shuttle mission. Crystal Eastman, who helped establish the American Civil Liberties Union, was also from Elmira. What a place.

WHEN I WAS a lad, high school students in Canada were set the task of learning all of the American state capitals. More than anything, it was probably a mental exercise. After all, I have successfully navigated the last thirty years of my life without once being called on to shout out in a bar that the capital of Vermont is Montpelier, or that in North Dakota legislators flock to Bismarck. At the time, it seemed to me that a lot of American state capitals didn't make a lot of sense.

They are often teeny places in states with a vast number of citizens. For instance, New York State has about 19 million citizens. Claiming just one-half of one percent of that population as residents, Albany somehow managed to become the state capital.

There must have been serious competition among candidates when it came time to choose each state capital. With the title comes glory. For instance, Albany can boast a major convention center, one of forty-seven campuses of the State University of New York, a new and exciting solid waste management landfill expansion project, and a truancy abatement project that has increased school attendance by 18 percent while reducing daytime juvenile crime by 13 percent. Even though it is a small community, the capital's rich social calendar includes festivals of tulips, storytelling, dance, Iroquois art, songs, blues, food, lobster, apples, strawberry shortcake, wine, nations, Celtic, Latin, Italian, jazz, and Shaker crafts. And surely that would be enough festivity for anyone. But then you would find that Albany is also home to the New York State Museum, and the museum harbors two stuffed Labrador Ducks.

At the museum, Jane went for a stroll while I got on with my ducks. She found that downtown Albany is a pretty sleepy place. Sleepy? Oh, come on, Jane! Albany boasts no fewer than eight farmers markets and a science fiction and fantasy fan club whose members are all gay, lesbian, or bisexual. And yet Jane found that if she had wanted a sandwich, a wedding dress, meditation crystals, or a tattoo, none were available before noon. Her only option was a coffee shop in which she read a novel.

Late in 1958, Paul Hahn received a polite letter from the New York State Museum in response to his questionnaire about stuffed specimens of extinct birds. Written by E. M. Reilly Jr., the museum's senior curator of zoology, the letter rambled on and on about the museum's holdings of Carolina Parakeets, Eskimo Curlews, Passenger Pigeons, and Heath Hens. He explained that this bird had been collected at Cranberry Island in 1886, and that that one had been collected in north Saskatchewan in 1896. Reilly then stuck in a brief note about Labrador Ducks. He explained that the museum had a pair, but nothing was known about them. Pretty typical. Between them, they would increase my total by two.

The museum's collection of birds isn't huge, and the Labrador Ducks are among its very best items. But even the ducks are exceeded in wow value by the Cohoes Mastodont, an immature male whose bones were unearthed in 1866. Radiocarbon dating showed that he had died 11,079 years earlier. In April. It was a Saturday. Not a particularly good Saturday from the perspective of the mastodont.

Labrador Ducks 48 and 49

Joe Bopp, collections manager of birds and mammals, had been with the museum for seventeen years. He was sporting short hair and a big black beard, which I suspect he used to cover up a big bubbly smile that he couldn't have removed with a scouring pad. He sat me down in front of my next two Labrador Ducks, an adult male and a reportedly adult female, both taxidermic mounts. As mentioned, not much is known about their provenance. They were already in the State Museum when someone got around to writing reports in the late 1840s. They are thought to have been shot somewhere around Long Island, sometime around 1840. The hen, a uniform cinnamon brown with a slightly darker back than belly, was mounted very close to her base, as though trying not to be seen. Only two of her tail feathers remained. While other specimens had brown, yellow, or red glass eyes, hers were clear glass, relieved only by the black pupils. The drake was far jauntier, looking as though he were about to be fed. His eyes were gray. In both cases, a preparator had been a little too enthusiastic with the paint pot, coloring their bills with blobs and splotches of black, mustard yellow, and gray-green. Both birds stood on simple white wooden bases; the underside of his held the words "Labrador Duck male De Rhem Coll. Presented 1850," and hers read "already in museum 1847 ♀ Labrador Duck see State Museum Report 1848." With these specimens behind me, the end of my quest was in sight.

BOSTON HAD BEEN really messing me around when it came to figuring out just how many stuffed Labrador Ducks remained. Across the Charles River, in Cambridge, the holdings of Harvard's Museum of Comparative Zoology seemed pretty straightforward. According to Hahn's list, Harvard had three ducks. The first was a female shot in

Nova Scotia in 1857. The second was an adult male from the collection of someone named Thayer. The third was an adult male with no details about where he came from or when. Great. But was there another Labrador Duck lurking in Boston? No such duck was listed in Hahn's book. I had been told by very reputable authorities that no such duck existed. Forget about it. Go for lunch. Drink a beer. Hmmm. But I had come to distrust very reputable authorities in the same way that I distrusted flamenco dancers, and I wasn't going for a beer until I had nailed this duck down.

In 1891, William Dutcher wrote an article about Labrador Ducks in the scholarly journal *The Auk*. In it he quoted Charles B. Cory, who said that the collection of the Boston Society of Natural History included an immature male Labrador Duck. Theodore Lyman had donated it to the society years before. It was thought to have been shot on the coast of New England, but no specific date or locality was available. I have already indicated that the immature male in Frankfurt had made its way from Boston via the American Museum of Natural History in New York. But a careful reading of the literature showed that the Frankfurt duck and the duck donated by Lyman were different specimens.

I live in fear that my book will land on bookshop shelves and ten minutes later someone will discover a stuffed Labrador Duck that I hadn't known about. In order to try to avoid the embarrassment that such a discovery would produce, in the days leading up to my American expedition, I scrambled after a fourth Labrador Duck at the Boston Society of Natural History. It didn't help that there was no such thing as the Boston Society of Natural History.

Or at least there wasn't anymore. One hundred seventy-five years earlier, six Bostonians had the foresight to establish just such a society. For more than three decades the group promoted the collection and study of all things related to natural history, working out of temporary quarters until a permanent facility could be constructed in the Back Bay region. After World War II, the city negotiated a ninety-nine-year lease for land in the area now known as Science Park. The nice thing about ninety-nine-year leases is that no matter how badly you screw up on the original deal, you are guaranteed to be dead by the time it has to be renegotiated. Sort of like Hong Kong, I suppose.

Along the way, Boston's natural history group had metamorphosed into the Boston Museum of Science.

Did the Boston Museum of Science have a Labrador Duck? Time was running out. In the nick of time an email message came in from curator Shana Hawrylchak. The museum did indeed have a stuffed Labrador Duck, she wrote. It was on public display, but if I wanted to examine it, she would be pleased to pull it out of its cabinet. What a sweet human being. She had brought the global tally of Labrador Ducks to fifty-five. One for each playing card in a deck, plus the jokers, plus the card of instructions for counting points in bridge.

Amtrak dropped Jane and me at Boston's Back Bay train station, and we took a cab to Cambridge, on the north side of the river. It was too late in the day to see much of the city but early enough for some power drinking. The hotel's shuttle bus took us to Harvard Square, which looked a lot more like a triangle than any other geometric shape. Being St. Patrick's Day, we expected all of the good bars to be crowded and all of the bad bars to be absolutely packed. We settled into a promising-sounding placed called the John Harvard Brewing Company, designed as an English-style pub. Clearly the architect had never been to England. Or sat in a pub. Few English pubs seat 300 patrons on high stools at long raised benches. Even fewer have stained-glass windows depicting saints. I didn't recognize all of the saints, but I did make out the late President Richard Nixon, author Ken Kesey, hockey legend Bobby Orr, and feminist Germaine Greer.

This was clearly a student hangout, and the saints were probably Harvard University alumni, or Harvard wannabes. While getting in the first round, I met a Harvard MBA student who admitted that he was likely to become filthy stinking rich. He also admitted that the place was usually this crowded, except on Fridays, when it was much worse. Pity the poor serving staff; my ears were ringing from the background noise. Jane and I started to wade through the long list of specialty beers. We had Frostbite Lager and Irish Red Ale and Demon Double Pale Ale and Brimstone Red. Jane took swigs from my glass without asking. She sneaked French fries off the plate of a neighboring patron each time he turned his head.

We spied on a nervous couple at the next table who were carefully avoiding looking at each other, like a mated pair of kittiwakes nesting on a narrow cliff ledge. When we lost interest in them, we watched a tangle of four young ladies. The lady in a mushroom-colored coat, checking out the action at other tables, slowly disentangled herself from her trio of companions. Short brown jacket tried coming on to white sweater. White sweater seemed oblivious to the advance, and so brown jacket moved a lot closer. Long black leather jacket would be dead ten years before realizing that some of her college friends were lesbians.

When we lost interest in that group, we chatted with Dustin Hoffman and Fred Flintstone, who were seated across the table from us. They were finishing off some sort of business diploma at Harvard. "When your boss tells you you're going on an expenses-paid trip to Cambridge, you don't say no," Dustin explained. They correctly guessed Jane's accent as "somewhere in Europe." Dustin couldn't place my accent, but his rather tipsier friend Fred got my accent as "north of the border." When we lost interest in them, we drank more beer.

IT WAS TO be a big day. While Jane had a little lie-in, I prepared to swoop down on Harvard and the Museum of Science to snap up four ducks. If everything went according to plan, I would increase my total of Labrador Ducks seen to fifty-three, and decrease my total of unseen ducks to two. If everything went according to plan . . . With duck-detection kit in hand, I set off in search of wisdom and truth.

Random wandering brought me to the Harvard Museum of Natural History. Plunked down right in the middle of the august campus, it is open to the public from 9 to 5 daily. Admission is very reasonable and anyone with Harvard I.D. is admitted free, although it seems to me that if you can afford to attend Harvard University, you can probably afford the price of admission to the museum. The research collection of animal bits and pieces goes by the rather grand name of the Museum of Comparative Zoology, Harvard University. Louis Agassiz, a brilliant zoologist and the museum's first director, founded the institution in 1859.

Alison Pirie, Collections Assistant in the museum's Department of Ornithology, met me in the museum's lobby and took me to the zoology research collection. In the elevator to the third floor, I told

Pirie how keen I was to see the collection's three ducks. "Two ducks," she corrected. I thought about this for a minute, and said, "I'm pretty sure you have three ducks." "Well, yes," was her reply. "If you count the one on display." Oh, God, please. Not again. Please, please, please tell me that I will get to examine the duck on display. Coolly, I said, "It won't be a problem to examine the third duck, will it, Alison?"

I was told that the fellow with the key, Ed Hack, was just back from holiday in Hawaii, and was too busy that morning to pull the Labrador Duck from its display. "How about if I come back to the museum after lunch?" Pirie put in a call. Yes, it seemed that arrangement would be satisfactory to Hack. Sight unseen, I started to dislike Hack.

So I settled in to work on the two ducks immediately available to me. Unlike most museum workspaces, this room was flooded with light pouring in through large arched windows. A space had been cleared for me at a bench.

Labrador Ducks 50 and 51

These Labrador Ducks are both study skins. They were listed as a hen and an adult drake. The female certainly seemed to be all of that, and I felt a little bad that she seemed dull to me. Nothing too very special about her; she was collected, by persons unknown, while migrating through Nova Scotia sometime around 1857. At the time of her demise she was kind of dull brown and slatey gray. Her legs, toes, and toenails are all very dark brown. One of the three tags around her legs indicated that she was "—very rare—" but that didn't keep her from being rather dull. She had been given a little too much stuffing, but these things happen. The only peculiar thing about her was a piece of wire poking out of her forehead, just above her bill.

The male was a bit more interesting to look at. Although listed as an adult, he clearly wasn't. His neck ring was dark brown instead of black, and his cheeks were gray instead of white. At the time of his demise, he must have been one molt away from adulthood. His bill, unpainted, sported colors of all the most valuable woods. The records I had been provided with said that nothing was known about his history. The tags around his legs were a little more illuminating.

Apparently, he had been in the collection of George Warren of Troy, New York, and was then bought by Dr. Thomas Heimstrut, also of Troy, and subsequently found his way into the collection of William Brewster before settling in at Harvard.

Harvard University doesn't sleep, and construction of a new building was under way next door. Workers must have been driving pilings, because about once per minute the whole museum shook. With each crash, the duck I was working on made a small leap off the desk. Being a jumpy sort of person, with each crash I made a somewhat bigger leap out of my seat. It was a bit like the scene in Jurassic Park with the steps of an approaching *Tyrannosaurus* making ripples in a glass of water, only worse. And with ducks.

After lunch I set off for the underground portion of Boston's public transit system, affectionately known as the T. I had been told that the Green Line would, if taken in the right direction, deposit me at Science Park, home of the Boston Museum of Science. Some of the route was a bit scary. Dimly lit offshoots of the main tunnels seemed to be the sorts of places where you might find a lost tribe of trolls. Construction on the line meant that I needed to take a shuttle bus to the museum.

The people of Boston are lucky enough to have two Labrador Ducks on display. They can see an adult drake up the road at Harvard, and a younger drake at the Museum of Science. I was met at the museum by Shana Hawrylchak, Collections Intern, and Curatorial Assistant Andy Grilz. Grilz is as big as Hawrylchak is slight. In the short time I was there, Grilz gave me the impression that he took joy in everything and pleasure in the company of everyone.

Labrador Duck 52

Hawrylchak and Grilz pulled the young bird out of the display cabinet that it shares with an Eskimo Curlew, a Passenger Pigeon, a Heath Hen, and an Ivory-billed Woodpecker. The duck is the star of that show. They set me up in a windowless workroom. Before departing, Grilz pointed out that the room had a closed-circuit televison camera. Not that he expected me to do anything wrong, but that I might want to avoid scratching myself in private places if I didn't want to end up on the blooper reel shown at the museum's Christmas party.

As was claimed when the duck was first noticed, well over a century earlier, hiding in the depths of the museum's collection, this male had come to the end his life while still in immature plumage. He was still brown in places that were destined to be black, and gray where he might have eventually become white. Someone had tried very hard to make this a pleasant specimen, although I suspect that the taxidermist had been handed a big job. The legs and toes had been treated with some thick touch-up compound, probably to cover damage. The proximal portion of his bill was painted a color that I had never seen before. Not just on ducks; I had never seen this color before on anything. It was light orange-yellow-pink. His glass eyes are hazel. Many of the specimens I had seen had badly worn tails but nearly immaculate wing feathers. In this case, the tail and wing feathers were both frayed and a few were broken. Even so, his posture was jaunty and his heavy plaster base had been painted to look like a seaside rock, gray and green as though coated with algae. What more could a dead duck ask for?

AND SO IT was back to the shuttle bus, and onward to the subway's Green Line with a transfer to the Red Line, and then off at Harvard Station for my last American duck. When I called for Ed Hack at the museum's reception desk, I expected to be met by some miserable old sod who couldn't be bothered to help me that morning. Instead, he struck me as the smaller, quieter member of the magician duo Penn and Teller. Penn, I think. Or maybe Teller. He had the smile of someone who had dropped off all his troubles in Hawaii and didn't mind being back at work at all. Hack had, in the interval, retrieved Harvard's third Labrador Duck from the display cabinet that he normally shares with an assortment of extinct and endangered birds, including a Passenger Pigeon, a Whooping Crane, a Great Auk, a California Condor, and one of only nineteen extant stuffed Guadalupe Island Caracaras. The cabinet may have been long on spectacular contents, but was rather short on style. Personally, I would have thought that Harvard would want to invest in some slightly nicer displays.

Labrador Duck 53

An adult drake at Harvard's Museum of Comparative Zoology, it is one of the few specimens on public display.

I found the duck to be a very pleasant-looking adult male, standing on an unmarked white wooden kidney-shaped platform, normally held to the backdrop by a metal L-bracket. His left side faces out toward visitors. His bill, legs, and feet had all been painted over in yellow, orange, and black, but I was getting rather accustomed to that. Whatever tags might have once adorned his legs had been discarded. If I screwed up my eyes, he seemed to be leaning forward ever so slightly, as though about to take flight. Not taking flight when he had the chance is probably what got him into the display cabinet.

THE DAY HAD left me beat. I dragged my duck-detection kit back to the hotel to wait for Jane to complete her day of touring. When she arrived, I admitted that if we didn't get going right away, I would probably crash and be lost to her for the night. After a pizza, we decided to spend the remainder of our last night in America back in the "English pub" close to the university. Not quite so crowded as the night before, it still had about 250 patrons. But for four of them at a table in the corner, I was the oldest person in the bar. I got the first round of drinks in to celebrate a very successful duck quest. Jane grabbed us a bit of table space and introduced herself to two attractive young ladies seated there. When I arrived, Jane introduced me as her friend "Greg." I reintroduced myself as Jane's friend "Glen." Honestly! You put a couple of pretty faces in front of some people, and they go all to pieces. The ladies were in the first year of university, studying Spanish, a program which they expected to finish in five years. I asked what one did with a degree in Spanish. "Teach Spanish," they said in unison. I was relieved that they hadn't asked me what one did with a degree in ornithology. I might have had to tell them the truth.

Chapter Sixteen

The Beano Goes to Russia

For a child in Britain, the most familiar comic book isn't *Fantastic Four, Archie,* or *Batman*. Instead, it's *The Beano,* and has been since 1938. Short on storyline but long on puns and characters playing naughty pranks, Roger the Dodger, Minnie the Minx, Dennis the Menace, and other chums with the middle name "the" have been encouraging young readers for the better part of seven decades. As a youngster in Canada, my family managed to find me the occasional copy of *The Beano*, possibly to promote a sense of my British Heritage.

A few years back, *The Beano* decided to get its young readers more involved in the comic by publishing their photographs, jokes, and drawings. The single best way to ensure that your picture appears on the Reader's Corner page is to have Mom or Dad snap a photo of you reading the comic while standing in front of some notable foreign monument, like the Sydney Harbour Bridge or Reactor 4 at Chernobyl. Darned and determined to get my photo in *The Beano,* when Lisa and I were ready to dash off to Russia to see my second last Labrador Duck, I made sure that I had a recent issue stashed in my luggage.

By all rights, St. Petersburg should be absolutely swarming with tourists. For those with a sense of culture, St. Petersburg offers more than three dozen museums, including the Bread Museum, the Artillery Museum, the Toy Museum, and the Museum of Hygiene. The more adventurous visitor from abroad need never feel homesick with bars and nightclubs like Saigon, Manhattan, Liverpool, Havana Club, Hollywood Nites, and Mollie's Irish Bar. The exchange rate of yen to roubles is a steal. All Sunday newspaper travel supplements have advertisements for package tours to Moscow and St. Petersburg, although none seem to feature ads for travel to sunny Novosibirsk or perky Yekaterinburg. People should be keen to travel to Russia if only because, until recently, it was almost impossible to do so—the place cries out for tourists.

But then comes the strange bit. Even at the height of the tourist season, Lisa and I found foreign visitors to be pretty scarce, and I think I know why. Most countries seem ever so keen to earn some extra cash by encouraging visitors from abroad. Yet, while America claims to be fighting a war on terrorism, Russia would appear to be fighting a war on tourism. In most countries, one need only show up with a credit card or some cash, a valid passport, and no more than a modest criminal record in order to be ushered in with open arms. Not so Russia. A visa must be applied for, specifying exact dates for arrival and departure. Don't even think about enjoying your stay enough to want to extend it a day or two.

After our passports had doubled in weight from all of the new glue-ons and staple-ins, the paperwork still continued on the flight from London's Heathrow to St. Petersburg. We were instructed to complete a registration card to get in, a registration card to get out, a customs declaration to get in, and a customs declaration to get out. The forms asked all the typical stupid questions, and a few more besides: "Are you carrying a gun? Are you importing a car? Are you exporting a car? Are you reexporting a car? Are you a descendant of the last czar intent on reestablishing imperial Russia through a bloody coup?" We wouldn't dare forget to register with the Visa Registration Department within three days of our arrival, unless we wanted a nasty surprise. Finally, we were not to pay the cab fare in anything other than roubles, even if roubles were the last currency our cabdriver wanted.

Our cabdriver, Ivan, looked as though he had, until recently, been a hockey player waiting for his big break into the National Hockey League. He sported broad shoulders, chiseled good looks, and a broken nose. His cab sported an air freshener with a picture of a naked woman and a tiger; neither seemed to be enjoying themselves. St. Petersburg can boast a little more than 4 million residents, making it more populous than Berlin or Los Angeles but slightly smaller than Calcutta or Wuhan. St. Petersburg has built out rather than up, and Ivan swept us 9 miles through the city's sprawling south end to our hotel. He had to wait in the lobby until we could trade in some dollars for roubles to pay the fare.

Our hotel was one of the big ones, with nearly 1,000 rooms, but it all looked a bit tattered. While registering, we were told that the tap water was perfectly safe for brushing our teeth, but that we weren't to swallow any of it. Having seen a bathtub full of it, I can see why—its hue was somewhere between Danube Canal green and dog wash brown.

St. Petersburg is a long way from anywhere. Except Helsinki, which is just 188 miles away. Since we hadn't flown in from Finland, Lisa and I were dragging our tails after a long day of travel. Dropping off our bags and grabbing a bite at the hotel seemed like the best option. I ordered a vegetarian pizza and a German beer; there was no Russian alcohol on offer. Lisa was a little more adventurous, ordering a *Brokkele Paste* off the English menu. Her meal had a remarkable resemblance to broccoli pasta.

Rejuvenated by the food, we found enough zip for a short walk along the Fontanka River. The canal-side avenue was quite choked with couples smoking, drinking, and snogging in almost equal measures. The waterway was choked with tour boats, although the boats were not choked with tourists. Despite the late hour, the sun was a long way off the horizon. By good fortune, we had arrived in late June. St. Petersburg is sufficiently far from the equator that at that time of year, the sky remained illuminated around the clock.

Looking back at our hotel, we had no trouble spotting our room; it was the only one with its window open. Odd, we thought. As we tried to drop off to sleep, it became apparent why all the other windows had been closed. Built on a big, fetid swamp, St. Petersburg is ideal breeding habitat for mosquitoes, and most of the year's bumper crop had found their way into our room. Every ten minutes, we had to get out of bed, turn on the lights, and try to swat the new pulse of mosquitoes that had snuck out from behind the curtains. This game of nude mosquito-bashing continued until about 2 a.m. Not that we had killed them all by then, we were just too tired to care anymore.

VIENNA IS OLD, Paris is antique, and London is practically fossilized. By comparison to other European centers, St. Petersburg is a bouncing baby boy. A slip of a lad himself, Peter became czar of Russia in 1682 at ten years of age. Peter wasn't particularly keen on the city of Moscow, possibly the result of having watched the bloody murder of his family there. Hence, he sailed down the Neva River to the Baltic Sea and set up shop in 1703. It can hardly have evaded Peter's notice that the area he had chosen for his new city was a great stinking, mosquito-infested swamp. Swamps don't make good building sites, and at least 40,000 people died trying to get the whole thing to stand up. Even so, by the time Peter died, in 1725, St. Petersburg was home to 40,000 residents. Strange numerical coincidence, there.

Peter the Great is credited with a pretty hefty range of rascally behaviors, including subordination of the church, subjugation of the peasantry, and the execution of anyone who ticked him off. It was also kind of sneaky to name his new city after a saint who shared his name. But whatever else Peter might have been accused of, you have to give him bonus points for his position on equality of the sexes. Initially, it might seem a bit extreme that William Mons was executed for sleeping with Peter's wife, Catherine, and that his pickled head was put in Catherine's bedchambers as a reminder. However, seeing both sides of the coin, Peter had his own lover, Mary Hamilton, executed and pickled as well. Peter's motto was apparently "I am one of those who are taught, and seek those who will teach me."

Despite the popular image of Russia as the new center of capitalism and everything nasty that goes with it, organized-crime bosses in Russia have bigger concerns than fleecing tourists. In St. Petersburg, tourists have less to fear from the mafia than being robbed and/or beaten up by thugs dressed as police officers. How comforting.

Indeed, there are really only six important threats to tourists to St. Petersburg:

1. Maniacal drivers have no respect for pedestrians. It is said that the majority of drivers do not have liability insurance—or brake pads.
2. St. Petersburg pickpockets display a remarkable level of industry. Curiously, German tourists are particularly frequently targeted. The police telephone hotline for foreigners is 278-30-18.
3. Missing manhole covers are an ongoing problem. I can't claim that we saw more than two dozen gaping manholes and managed to avoid every one.
4. Every guidebook warns of Romany beggars, while claiming not to be bigoted toward any particular ethnic group. Begging can quickly turn to flagrant theft when women work with children trained to rifle through pockets.
5. The drinking water is contaminated by heavy metals, bacteria, and a nasty intestinal parasite called *Giardia lamblia*. Having suffered through giardiasis in Canada, you can trust me—you don't want this one. Unless vomiting and explosive diarrhea are a priority for you.
6. The final great tourist problem is a national toilet paper shortage. The last two problems may be related. Carrying a small satchel of toilet paper is said to be the best way to go.

THE ZOOLOGICAL INSTITUTE was not scheduled to open until 11:00, which gave us time for a bit of a lie-in and a leisurely stroll along the Fontanka River and the Voznesenskiy Prospekt. This neighborhood isn't the core of the tourist map, but lies well within its southern limits. The community had the sense of loving neglect—a place where people get on with their lives. Nowhere were the buildings crying out for repair, but everything was in need of a good scrub. We passed workers removing pink and purple graffiti from a pillar, but no one was working on the general grime with a brush and bucket of soapy water.

After crossing a couple of canals, our aimless wandering brought us to St. Isaac's Cathedral, one of the biggest churches in the world. The engineers responsible for St. Isaac's should get big kudos for managing to erect it on swampland. Its gold dome is a great navigational landmark, visible for miles. Ironically, during the Soviet era, St. Isaac's was converted into a museum in celebration of atheism. Since my second-most-important task in St. Petersburg was to get my photograph published in *The Beano*, I brought out my camera and asked Lisa to snap a quick one of me holding up the comic with St. Isaac's in the background. I tried to look impish.

Passing through the park surrounding the Admiralty, we got our first look at the magnificent Neva River where it divides into the Bolshaya (Great) Neva and Malaya (Small) Neva. Crossing the river at Dvortsovy Most, the Palace Bridge, we arrived on Vasilevskiy Island, which Peter had intended to be the focal part of his new city. There were some initial problems with frequent flooding and lack of reliable river crossings, but Vasilevskiy Island did eventually catch up with the rest of the city.

We sat to gather our thoughts in the park across from the Naval Museum. Lisa noticed that the residents of St. Petersburg are not shy about making eye contact, but also not in a big hurry to return a smile. We were later told that this reluctance is a holdover from the Cold War era, when it was not at all clear who were your friends and who was likely to turn you over to the authorities.

In essence, the city's zoological collection was established by Peter the Great, just ten years after he established St. Petersburg. The Zoological Institute, formally inaugurated in 1832, consists of some 15 million specimens of 280,000 different animal species. According to the institution's website, 40,000 of these specimens are on exhibition in the public galleries. Putting many Western facilities to shame, the institute's library has more than 500,000 books and journals. The website also provides a delicious quote about St. Petersburg's Zoological Institute that seems an equal mixture of pride and propaganda. "Solidarity, devotion to their work, and enthusiasm characterize the staff of the Zoological Institute. These number approximately 500 people. As a rule, scientists and technicians work at the Institute for many years. For many of them, the Institute is their only work place during their lifetimes."

The magic appointment hour arrived and we walked to the entrance of the Zoological Museum and Institute. From the entranceway, it didn't seem sufficiently posh, and we wondered if we might be in the wrong place. For such a grand institution, the cubbyhole of an entrance seemed better suited for a junior high school in a small prairie town. As we entered we passed a group of students, no doubt about to get a tour, composed of fourteen bored-looking sixteen-year-old fellows, and one really excited-looking, estrogen-fueled sixteen-year-old girl. After they cleared the foyer, I called Dr. Vladimir Loskot on the first rotary telephone I had seen in about thirty years.

Loskot, Leading Research Fellow and Head of the Department of Ornithology at the institute, looked to be somewhere in his early sixties. This may be entirely unfair—ornithologists generally look quite a bit older than they are, having endured the ravages of sun and wind in the pursuit of birds. Loskot explained that his previous thirty years had been spent at the Zoological Institute. He led us through a level of public displays, through a security door, and into the troglodyte world of the institute's research collection. Underfoot, the concrete floor was crumbling in places and excavated in others, with some of the more dangerous bits covered with metal sheets. We passed a small mountain of rusting radiators that had been ripped out awaiting replacement, hopefully before winter set in. We spied a slightly smaller hillock of fluorescent tubes. Whether new or burned out, they were covered with a layer of dust thick enough to be called soil. Indeed, the whole tunnel system was dusty enough to support its own ecosystem. We passed a cleaning lady who, in the ultimate act of futility, was trying to clear away some of the dirt with a small brush. Loskot called the elevator. It was one of those remarkable optical illusions—much smaller on the inside than it appeared on the outside. With the three of us inside, there was enough air to fill only one pair of lungs, so we had to breathe in turns. In broken English, Loskot used one of his breaths to explain that we should avoid touching the elevator's walls as they were probably dirty.

Lisa and I settled into a workbench near a window, surrounded by storage cabinets that reached from the floor to the ceiling high above. These cabinets contained one of the world's great ornithological collections. It can boast about 170,000 skins of 4,200 bird species. It also has 2,700 skeletons of 1,080 species, 850 species represented by 7,500 alcohol specimens, and a jumbo collection of nests and eggs.

Labrador Duck 54

St. Petersburg's Labrador Duck, a drake, was something of an oddity. Although it was a long, skinny study skin rather than a taxidermic mount, it had light brown glass eyes. I suspect that this specimen started its life-after-death experience as a taxidermist's creation and for economy of space was later converted into a study skin. A really nice job had been done of it, and the specimen retained a touch of the artistic. There are some small pinholes in the webs between the toes, which suggests to me that it was pinned to a display board sometime in the past. It resides in a cardboard box, ornately decorated with the beautiful paper often found on the inside covers of precious old books. Between examinations, the Labrador Duck shares its box with an extinct Heath Hen.

When I first contacted Loskot ten years earlier, he explained that his duck had come to Russia from the Hamburg dealer G. A. Salmin sometime between 1830 and 1840, but the rest of its history is lost to the mists of time. For a spell after World War II, the hen from Dresden via Königstein joined the drake, but you have already read that story. When Paul Hahn constructed his book on extinct North American birds, he had been able to gather a lot less information on this specimen, perhaps because he was working under the restrictions of the Cold War. When his book on extinct birds came out in 1963, he was able to write only: "UNION OF SOVIET SOCIALIST REPUBLICS: 24. Musej Zoologicheskogo Instituta Akademii, Leningrad. Male adult."

I finished poking and measuring the duck and put him back in his box. After the gentlest of hints, Loskot generously offered to give Lisa and me a tour of the public galleries. They proved a most startling contrast between the wealth of their contents and the shabby, third-rate way in which they are displayed. On offer were three mammoths, one of which had been excavated from the frozen wastes of Siberia in 1902, some 44,000 years after it had died. Other displays included dioramas of penguins and northern seabirds, and a particularly good grouping of ungulates with corrugated noses.

Despite this wealth, the displays themselves were horrid. The lighting was poor, the wooden bases crude, paint flaked everywhere, and the cheap glass in the displays was rippled. Old cardboard labels read "CCCP," even though the Soviet Union had dissolved fourteen years earlier. By chance, we had arrived on the sixty-fourth anniversary of the start of the war between Russia and Germany. These hostilities were commemorated by the glass panels of a display, still perforated by bullet holes. Further along, a display of skeletons of whales and seals was oddly perforated by a giraffe skeleton. The whole area was in a state of repair, disassembly, construction, disintegration, or demolition; difficult to say which. Clear sight lines were a thing of legend. The floor was erupting everywhere, and if the museum were in America, it would be closed immediately for fear of a major lawsuit.

However, to describe the museum as a dump would have been uncharitable, callous, and a gross misrepresentation. This is a great museum that has the misfortune to be situated in a country that can't afford to keep it up to a high standard. The displays were much better attended than those in some of the far fancier museums I have seen in richer countries. The museum hosts between 700,000 and 900,000 visitors each year. On the day of our visit, several young people used the exhibits to practice their painting and sketching, with an albatross being rendered in ink and a petrel in pastels. Much like Russia herself, the Zoological Institute is stationed somewhere in the abyss between here and there.

When I had first contacted Loskot about my visit, I had offered to take him and his wife, Vera, out for a meal. On the day, they showed incredible hospitality by hosting lunch in the room outside Loskot's office. The room spoke of a day when natural history museum research was a priority. Cabinets were filled with great volumes of ornithological study. Whether they belonged to the institution or to Loskot wasn't clear. Beside the window perched an old record player of the same vintage as the telephone in the entrance foyer. Vera served us well-steeped tea which was then diluted with hot water, a Russian tradition. We were offered rye bread, ham, tarts, and chocolate-covered biscuits. As we ate, we chatted and discovered many common interests. We found that Vera studies the taxonomy of parasitoid flies, not so far from my interest in fleas. I found that Loskot and I share a passion for the song dialects of birds and jazz music. He showed me his directional microphone and digital recording machine, which, quite frankly, is a hell of a lot nicer than my dingy old analogue equipment. Lisa and I pulled out our Russian phrase book to try out a couple of the simpler expressions, but, regrettably, for those raised on English, some Russian syllables are nearly impossible to pronounce.

I pulled out a letter that Loskot had sent me ten years before in response to my inquiry about the duck in St. Petersburg. Vera described me as "dangerous" for keeping all my old correspondence. Could this reflexive caution be a holdover from the former Soviet era? Is caution still ingrained even in the most welcoming of people? With a twinkle in her eye, Vera also described me as "difficult" for turning down the offer of more food. Before we left the museum, Vera and Loskot warned us sternly about pickpockets and Romany thieves, and reminded us about the hazards of drinking tap water. After saying good-bye to Vera, Loskot guided us back through the dusty tunnel system to the museum entrance. He clasped my right hand firmly in both of his and wouldn't let go until he could think of the proper English phrase. He said, "Please write."

BACK OVER THE Neva River, we purchased bottles of drinking water from a street vendor, which were cheap by anyone's standards. We did as we had been told, checking that the bottle had an intact seal; cheap is one thing, but heavy-metal poisoning is another. In the warmth of midsummer, vendors like her might be the only thing keeping the residents of St. Petersburg alive.

We walked broad boulevards and thought the streets resembled something more North American than typically European. With many other pedestrians, we waited at a major intersection for the walk light to turn green. The automobile traffic was continuous. After a period, a uniformed man with a walkie-talkie descended from a kiosk to stop the traffic. We all started to cross. "*Nyet!*" he shouted, and waved through a police escort and a couple of limousines flying diplomatic flags. Having seen the entourage pass, we resumed our crossing along with fifty or sixty other pedestrians. Again he shouted,

"*Nyet!*," ordering us all back to the sidewalk so that the traffic could continue unimpeded, spoke into his walkie-talkie, and climbed back into his kiosk. Surely this abuse of privilege is the sort of thing of which revolutions are made.

Given Peter's particular choice of geography, St. Petersburg requires an extensive system of canals. We chose to proceed west along the Moyka Canal. Crossing one bridge after another, we were amazed at the number of tour boats that plied the canal. At one point, near a large yellow building, the Yusupov Palace, every passing boat seemed to provide its patrons with a commentary. We heard the spiel in Italian, French, German, and Russian, but not in English, and so we had to pull out our guidebook. In the early twentieth century not everyone was entirely pleased with the influence that peasant Grigory Rasputin had over the Russian court. With the promise of a party, Rasputin was lured to Yusupov Palace on the evening of December 17, 1916. There he was poisoned. Leaving nothing to chance, Prince Yusupov then shot Rasputin. Checking back later, Yusupov found that Rasputin was still alive and after a brief struggle shot him three more times. Not satisfied with that, Yusupov had Rasputin's corpse beaten and tossed into the river. An autopsy, performed after his body was discovered three days later, showed Rasputin was still alive when dumped in the river and that he eventually died of drowning. Out came *The Beano,* and Lisa took another photograph.

Cutting south along the Kryukov Canal, peeping at every exotic building on offer, we found that the drone of tour-boat commentary resumed in front of St. Nicholas' Cathedral. Built around the same time as Yusupov Palace, the cathedral is a beautiful structure of pale blue and white, topped by golden domes. At St. Nicholas', it all came down to weddings, christenings, and other religious services—pretty typical cathedral stuff. The guidebook didn't mention anyone famous being poisoned, shot, beaten or drowned. Certainly very pretty; the camera came out again so Lisa could snap another picture of me and *The Beano*.

THE NEXT MORNING, arriving at the square around the Hermitage and Winter Palace at 10:00, thirty minutes too early for the opening, we watched workers rolling up acres of sod that had been put down the day before as part of a commemoration of the Day of Remembrance and Sorrow, when Hitler's forces had attacked. I brought out my camera and had Lisa snap a photo of me holding up *The Beano* in front of the Winter Palace.

First we queued to enter the Hermitage, and then found ourselves in the line to buy tickets. It became clear pretty quickly that Lisa and I did not have the proper temperament to deal with Russian line-ups. In St. Petersburg, stature and pushiness are inversely proportional. At the head of line I held up two fingers and tried a gruff Russian accent on the word "два," but I think it was the wrong genitive singular, and so had to pay the foreigner's rate—three times as much as for small pushy Russians.

In contrast to the austerity of the Zoology Institute, the Hermitage had weathered Russia's stormy transition to a free-market economy. Its floors were inlaid with the finest woods in intricate geometric patterns, and skirting boards were done in marble. Galleries featured great towering ceilings with high windows that allowed the most subtle filtered light to pass through. Walls were finished in red or blue or cream or gilt. Great urns and bowls had been created of marble, malachite, and jasper, the result of a competition to see who could use more construction material.

I discovered an interesting painting by Matthias Withoos (1627–1703) that featured a hedgehog, a frog, and two mice partying under a thistle. Interesting, but not particularly good. Paulus Potter (1625–1654) had chosen to paint the backside of a bull, although I cannot imagine why. Joachim Wtewael (1566–1638) had created a piece called *Lot and His Daughters*. The quality was good, but it left me asking why Lot's daughters were naked, and why Lot was copping a feel. Michelangelo's sculpture *Crouching Boy* may be the only piece by the artist in Russia, but it isn't a particularly good piece, having been knocked off while Mike was waiting for the pub to open. For all of the Hermitage's fame, Lisa felt the housing outpaced the contents.

After a lunch of sandwiches and ice cream, Lisa and I headed off down Nevsky Prospekt, St. Petersburg's main artery. Once named the Street of Tolerance, it is a promenade of churches, palaces, businesses, and retail opportunities. A must-see for any visitor, the avenue runs southeast from the city's core, where all well-dressed and soon-to-be-well-dressed citizens stroll, and where pickpockets are drawn to tourists like a pride of lions to a zebra with a hangover.

A left turn off Nevsky Prospekt at the Kanal Griboedova revealed the quintessential image of Russia. It was the church I would paint if I had any talent as a painter. And some paint. The Church of Our Savior on the Spilled Blood was constructed on the spot where Czar Alexander II was assassinated in 1831. A riot of mosaic enamel tiles and swirling domes and carved marble and gaudy colors, the ornamentation emphasized the pastel tone of the remainder of the city. The church also seemed as much a tribute to the nation's mourning over the loss of Alexander as a celebration of Christianity. Out came the camera, and Lisa snapped another quick one of me and *The Beano*. Hawkers around the church offered quick sketches, bottled water, toilet opportunities, and nested Russian dolls. Not all of these dolls followed the theme of traditionally garbed women, instead featuring Bill Clinton, Vladimir Putin, Harry Potter, and even Michael Jackson, although he was labeled "John Lennon."

SOME WORDS TRANSLATE from Russian to English fairly easily, but most do not. On a menu, for instance, you might be willing to guess that ordering the Омлет would result in the eventual appearance of an omelet, and you wouldn't be disappointed—unless you really didn't like omelets. Similarly, it doesn't take a rocket surgeon to squeeze the words *Coca-Cola* out of Кока-Кола. However, no amount of hard squinting could turn the word Шпинат into "spinach" as my phrase book suggested.

The prime difficulty isn't so much that Russian is a foreign language, as that it uses an alphabet that I just couldn't wrap my noodle around. Invented in the ninth century by a monk named Cyril, Cyrillic has, at least in a nominal sense, only seven more letters than English, but these include Г, Ж, and Я, none of which are pronounced like they look. On top of that, Cyrillic includes a bunch of diphthongs, like ЮЙ. Regrettably, after you learn a few of the simpler letters, you find that vowels and diphthongs sometimes, but not always, change pronunciation depending on whether or not they are stressed. "Stress in Russian," claimed my phrase book, "is irregular." And after all that, Russian is one of only a hundred languages in Russia. We prayed that we didn't get into a cab with a driver who spoke only Nivkh or Vespian.

My Russian phrase book was remarkably frank in the section entitled "Making Friends." It included some of my favorite expressions, including Ты прекрасо выглялишь! (You look hot!), Вам принести что-нибудь пить? (Can I get you a drink?), Можно тебя поцедовать? (May I kiss you?), Хотите пойти v отель (Let's go back to your hotel), and Можно с другом? (May I bring a friend?). If you can't have fun in St. Petersburg on a Friday night with those five expressions, it's time to check your pulse. Of course, that sort of fun should always be mixed with a degree of caution. Russia can claim two unfortunate statistics. It is at, or near the top of, the world's chart in terms of an increasing incidence of HIV. It also has a terribly high number of huge boyfriends with no sense of humor.

In late June the sun shines on St. Petersburg almost continuously. With all that warmth and the city's proximity to the sea, it isn't surprising that St. Petersburg is sultry around the clock. If I wanted to go running, the only real option was the early morning. The community moves at a very different pace at 6 a.m. Most of my fellow pedestrians were still up, not already up. I ran upriver, past a series of old, refurbished agricultural tractors towing water tanks to wash down the streets. I passed old, refurbished army trucks, now fitted with brushes to whisk up the newly washed rubbish. I crossed paths with a young lady in a short red skirt, vodka bottle in hand, dancing and thrusting to a song that only she could hear. Several young men tended fishing poles. It made me wonder how edible the fish would be if even the drinking water was contaminated with heavy metals. An older fellow in a 1970s-style Adidas track suit ran by on the opposite side of the street, but I spotted him only as he stopped waving at me. I then passed another runner, and squeezed a smile out of her by saying "Доброе утро," pronouncing it "Dobraye ootra." By the end of my forty-five-minute run, the homeward-bound populace had been largely replaced by the workplace-bound.

Even though the city is encased in ice for a fair swack of the year, the Neva receives its fair share of ships. To allow shipping during the clement months, all bridges across the Neva are drawbridges, opening in the wee small hours. No small number of homeward-bound party-goers find themselves on the wrong side of a bridge. We took to Vasilevskiy Island by crossing the Neva via the Most Leytenanta Shmidta, named after a sailor who led some uprising or other. Peter's original plan was to make Vasilevskiy Island the administrative center of the whole shebang. This didn't work out as well as hoped because of the island's nasty habit of flooding, and the difficulty of building stable bridges to it. On the island's south shore, just in front of the Academy of Arts, we found two great bronzy sphinxes, dating from the fourteenth century BC. Lisa, a fan of Egyptology, was a little ticked that she couldn't figure out which pharaoh they were based on. Regardless, they must be a big hit on the organized coach tours, as throngs poured out to photograph them. It didn't seem like a *Beano* moment.

From there we took a zip around the well-treed grounds of St. Petersburg's university, always trying to keep to the shade and ever vigilant for a bit of breeze. We crossed the Birzhevoy Most, and then the Kronverskiy Most, finally arriving on the Island of Nearly Naked Women, also known as the Fortress of Peter and Paul. Lying on the beach along the island's south side and nestled into every patch of long grass were delightful examples of the female form, most of them in advanced stages of undress. I didn't spot a single lady who had removed absolutely all of her clothing, although I might have tried harder if Lisa hadn't been with me. Oh, yes, and there were also a lot of nearly naked men.

To be flippant about all the exposed flesh is to largely miss the point of the Fortress of Peter and Paul. For all intents and purposes, its initiation in 1703 by Peter the Great represented the founding of St. Petersburg. Many workers died in its construction and many political prisoners were interrogated, sentenced, and tortured within its walls. Despite being considered one of St. Petersburg's great literary heroes today, even Dostoevsky spent some very unpleasant time there. Circling the south walls, we took to the fortress through Neva Gate, also known as the Death Gate. Prisoners leaving the fortress through this gate were on their way to exile or execution, which frequently amounted to the same thing.

Visitors are welcome to wander the island for free, but to take in the myriad exhibits requires a single all-inclusive ticket. When Russia's war on tourism is over, it will do well to end the practice of charging foreign visitors much more than residents. Again, admission was three times as costly for foreigners. At the ticket window, I tried to sneak in for the lower price by looking dour and holding up two fingers rather than saying "два." It didn't work any better than it had at the Hermitage, and we had to pay the full 240 roubles. Lisa found that only 10 roubles was required for access to a toilet with no toilet seat but that provided unlimited toilet paper. This is much better than the deal offered by most of the city's portable toilet operators, who charge 20 roubles and dispense paper by the square.

Just nine years after founding the fortress, Peter set architects to work on the Cathedral of Saints Peter and Paul. For the next 250 years, the cathedral's spire was the tallest structure in St. Petersburg. I suppose that we all develop an image of what a church should look like. This wasn't mine, and a small voice told me that it might not be God's, either. The ceiling was painted with innumerable cherubs, and I would be willing to bet my next paycheck that God doesn't like cherubs any more than I do. The cathedral, with way too much gilt on the outside, and way too many crystal chandeliers on the inside, is the resting place of most of Russia's czars. For most of those interred, the sarcophagi were comparatively simple, just big blocks of white marble. But not everything is as it appears. Soaring pastel marble columns turned out to be simpler stone, painted to look like marble. A small army of attendants tut-tutted to dissuade those who might touch, and a few uniformed officers sported holsters, which, presumably, contained guns. Big on gilt and chandeliers, but not so very big on God, I was rather surprised to cross paths with a priest.

The juxtaposition of the cathedral and the adjacent Commandant's House was unsettling. The Commandant's House, constructed just a few years after completion of the cathedral, served as the residence of the fortress's commanding officer as well as the site of interrogation and sentencing of prisoners. Although it was one of the spots allowed by our tickets, I found myself unable to visit the Trubetskoy Bastion Prison, last stop for prisoners before the Neva Gate. Peter's own son, Alexis, was one of the first political prisoners to be detained there. Poor sod. History tells us that after a period of interrogation on trumped-up charges of treason, Alexis managed to escape, only to be fooled into returning by the promise of a pardon from his father, who had him tortured and beaten to death. Indeed, throughout my visit to the fortress, I could feel the souls of those who had died constructing it and of those who had died evil deaths afterward. Lisa and I walked slowly and spoke quietly; my camera stayed in its bag.

Leaving the fortress, we recrossed the Neva, and found our way to Decembrists' Square, the spot where Russia's first revolution was quashed on December 14, 1825. At the inauguration of Nicholas I, rebel forces, intent on installing a constitutional monarchy, were routed. Some of the rebels were executed and a whole lot more were exiled to Siberia. The one unavoidable feature of the park was the huge statue of Peter on horseback, dressed in a Roman toga and capped with a laurel wreath. Peter, not the horse. The horse was busy stamping on a very large snake that somehow represented treason. The statue, commissioned by Peter's wife, Catherine, and executed by sculptor Étienne Falconet, was erected in 1782. For reasons not explained in the book, Peter's right arm was extended in a level salute. Lisa figured it out. "It's like an amusement park ride. He's saying, 'You must be at least this tall to forge an empire!' "

The park had its gentle side that afternoon. When a young Russian girl dreams of her wedding day, she must envision an early Friday afternoon in late June. She must imagine a silver-covered book full of photos of her fairy-tale princess dress, with the trees of Decembrists' Square as a backdrop. The park that afternoon was stuffed full of brides. With a quick scan we spotted twelve brides in white and one in hot pink. Thirteen grooms must have been somewhere in the park, but, being almost irrelevant on the big day, they blended in with all the well-wishers in gray suits. This was a day about blushing brides. Flashes popped everywhere, and I pulled out my camera, wanting to be part of the revelry. I wanted to see how many brides I could get into a single photograph. No matter where I stood or how I twisted, four was the best I could manage. I briefly contemplated trying to get a bride to pose with me and *The Beano* but decided against it. I was having enough trouble with simple expressions like "Где ближайший отделение милиции?"

Only a fool would claim that Russia has beaten back all of its demons. Her system of governance could use some tweaking, the economy is still shaky, and the environment is decidedly crappy in places. However, when a nation is blessed with a citizenry with an indomitable will, a wealth of natural resources, a vision of greatness, and a spirit that can shove thirteen brides into one small park, success cannot be far off.

• • •

Glen Chilton,
Dept. Biology,
St. Mary's University College,
14500 Bannister Road, SE,
Calgary, Alberta,
Canada
T2X 1Z4

Dear Glen,

CONGRATULATIONS! *I am delighted to say the entry you sent to the BEANO CLUB page has been featured in the comic. This means that you have won* **FREE** *Bronze Membership to the world-renowned BEANO CLUB!*

So that we can enrol you in the club, we'd like you to simply fill in the accompanying form and send it back to us in the return envelope provided. Remember—you don't need a stamp.

Once we have received the filled-in form, we'll do the rest and send you your brilliant BEANO CLUB goodies.

Well done again and welcome to **THE BEANO CLUB.**

Best wishes,
Euan Kerr

Chapter Seventeen

The Curse of the Labrador Duck

Some endeavors are doomed from the start. Think of the Tower of Babel, Paul Hahn's attempts to find every stuffed Passenger Pigeon in the world, or my attempts to learn organic chemistry. Similarly, some people can never figure out when to give up, and admit that an endeavor is well and truly doomed. Think of Captain Ahab and his blood lust for Moby-Dick, or attempts to convince tourists to visit sunny Mirmansk.

One of the greatest doomed fictional endeavors is that of Caspar Gutman in the novel *The Maltese Falcon*, by Dashiell Hammett. If you have seen the film version of the story, starring Humphrey Bogart, you will know what I mean. If you haven't, I'll save you the trouble of having to rush out to the video store by giving you the gist of the plot, without giving away the ending.

A beautiful young woman hires detectives Miles Archer and Sam Spade to find her runaway sister. The story about the sister is a ruse and Archer is shot dead. The young woman is one of several characters who will stop at nothing to find a priceless jewel-encrusted statuette of a falcon. Sam Spade becomes entangled with these characters, each hoping to make a lot of money when the falcon is turned over to Caspar Gutman, who has been searching for it for seventeen years. Death and mayhem continue.

In the search for the world's stuffed and now extinct Labrador Ducks, a few specimens are probably gone forever. But one specimen hung on the periphery of my vision like some mysterious, elusive, hanging duck-shaped thingy. The story involves a long and horrible chain of dead protagonists. Indeed, there are so many corpses that you may need a score card.

In 1952, rather late in the day for Labrador Ducks, William E. Glegg of the Zoological Museum in Tring published, in the prestigious ornithological journal *The Auk*, an article entitled "Discovery of an Unrecorded, Mounted Male Specimen of the Labrador Duck, *Camptorhynchus labradorius*." As the first new Labrador Duck to be discovered in over half a century, and indeed the last new specimen to be discovered until my quest began, this was quite a find. In his brief article, Glegg wrote:

> About 1947–1948 it was reported that another specimen of the Labrador Duck had been discovered in England. Mr. R.L.E. Ford of Messers. Watkins and Doncaster, London, who discovered and purchased this bird, informs me that it was mounted and a perfect drake. The specimen was found in a case of various mounted birds of North America in a country house where it had been for about 100 years, but nothing was known of its history. . . . Mr. Ford sold it to a private collector in Britain, but he is not at liberty to disclose his name or the price, which was substantial. Information has reached me from other sources that the specimen was found in Kent and was sold to Capt. Vivian Hewitt for 500 pounds, after being offered to the British Museum.

The next mention of this duck comes in Paul Hahn's 1963 book, *Where Is That Vanished Bird?* Hahn listed the birds by country in alphabetical order, and so after the two birds in Austria came bird number three. Hahn wrote: "BAHAMAS 3. Captain Vivian Hewitt, Derby Island. Male. See William E. Glegg, Auk, 1952 . . . ," after which he recounted some of the details provided by Glegg's article.

And that, as they say, is pretty much that. Where does one start, particularly since there is no such place as Derby Island, Bahamas? If you were me, you would write to the Canadian Consulate in Nassau, asking for help. After all, why do I pay taxes if not to occupy the precious time of Canadian diplomats in foreign countries? Mrs. Monique L. Brooks of the consulate wrote back to suggest that I get in touch with Mr. J. F. Bethell, President of the Bahamas National Trust. I did.

While waiting for a response from the National Trust in Nassau, I attended the American Ornithologists' Union annual meetings in Cincinnati, presenting a paper about the songs of White-crowned Sparrows, and another about fleas on Song Sparrows. Folks at the Cincinnati Museum of Natural History had prepared an exhibition of stuffed specimens of extinct birds. Among them was a Great Auk with a notice claiming, "The specimen and egg exhibited here were owned by the late Captain Vivian Hewitt and were sold to the Museum by Spink & Son, Ltd of London on 24 September 1974." This was my first breakthrough. Not only did I know that Captain Hewitt was dead, but I also knew who had handled his affairs. I raced off to find Bob Kennedy, Deputy Director for Collections and Research at the Cincinnati Museum, to ask if they had also purchased the Captain's Labrador Duck. He said they had not, but was able to provide me with the address of Spink & Son, antique dealers.

Spink & Son Ltd, must be quite the group; their letterhead suggests they work by appointment to Her Majesty the Queen, His Royal Highness the Duke of Edinburgh, and His Royal Highness the Prince of Wales. Natasha de Wiart of Spink & Son passed on to me the address of Miss Patricia McCawley, now retired, who had been the secretary of the late Mr. David Spink at the time when they handled Captain Hewitt's estate. Miss McCawley wrote to say that she and David Spink had handled all of the Captain's collections, including stuffed Great Auks and some auk eggs, but no specimens of the Labrador Duck. She went on to tell me the Captain must have sold the duck privately before his death, as it certainly wasn't among his effects listed by her and Mr. Spink in Nassau before their transport to London. In later correspondence, descendants of Hewitt's housekeeper, Mrs. E. M. Parry, fondly described Miss McCawley as a family friend.

What could I say about this duck so far? Captain Vivian Hewitt, now dead, is rumored to have purchased a Labrador Duck in England in the late 1940s for £500. He later moved to the Bahamas. When he died, Hewitt's collection was sold by the firm of Spink & Son, who have no record of the Labrador Duck. Spink is now dead, and his secretary knew nothing of the duck. So far, I really hadn't much to go on.

A few months passed, so I wrote again to the Bahamas National Trust. This time Mrs. Lynn Gape responded on behalf of Mr. Bethell, explaining that they had no information on either Captain Hewitt or his Labrador Duck, but that they had graciously run an advertisement in the local newspaper on my behalf. It read: "Anyone having information regarding a Captain Vivian Hewitt of Derby Island, The Bahamas, should please contact Dr. Glen Chilton . . . ," with my contact details.

Less than a week later, I arrived at my office to find my telephone ringing. A gentleman in the Bahamas, whose name I didn't catch, explained that he had been a colleague of Hewitt's in the Bahamas before the Captain took ill and returned to England for treatment. The caller asked if I knew that Captain Hewitt was dead and I responded that I did. He then asked me if I knew that the Captain had never married, and I replied that was new information to me. After a lengthy and uncomfortable pause, he said that Hewitt (ahem, ahem) may have left behind at least one illegitimate child (ahem, ahem), and that I might want to be in touch with Major Paul Kenneth Parry (ahem, ahem). Parry . . . Parry . . . Wasn't that the name of Hewitt's housekeeper? Apparently it was Major Parry who had directed Spink & Son to deal with Hewitt's substantial collections. The last my informant had heard, Major Parry had been living on Guernsey in the Channel Islands.

What in the world was I getting myself into?

Using my directory to members of the American Ornithologists' Union, I found an email address for Chris Clark in Jersey, not ever so far from Guernsey. Chris graciously snooped around. A couple of days later he told me that although Paul Kenneth Parry had passed away eighteen months earlier, he could provide me with the address of his widow.

My letter to Mrs. Parry was answered by her son-in-law, John Lipscombe, who explained that about 1965 Major Parry had directed Spink & Son to deal with Captain Hewitt's collection. John went on to explain that Major Parry had a surviving brother. "But I fear that he will not be able to help you much in your search." John was also kind enough to pass along my letter to Major Parry's nephew, Mr. Vivian Davies, who was living in Dorset, in the hopes that he might have some information about the Labrador Duck. In a follow-up letter, Mr. Lipscombe wrote: "Vivian Hewitt was not an organized man and I believe that he was not that with it for quite some time before he died."

This is a really cooperative family! Vivian Davies, the nephew of Major Parry, wrote to pass along what he knew of the later stages of Vivian Hewitt's life, and of his collection of natural history artifacts. He had spoken to his elderly uncle, the surviving brother of Major Parry, "who is a recluse of well over 80, living in North Wales, the only surviving member of my immediate family, who cannot remember any details concerning the duck or its disposal." Mr. Davies gave me two further potentially valuable leads. He put me on to a biography of Captain Hewitt, *Modest Millionaire*, by William Hywel, and described it as "an interesting account of a remarkable man." He also told me that, at the time of Hewitt's death, all of the bird collections were sent to Dr. Jim Flegg, director of the British Trust for Ornithology at Tring.

The current Director of the BTO, Dr. Jeremy Greenwood, responded to my letter of inquiry. He had asked several former members of the BTO staff for information, but none could recall coming across a Labrador Duck, and they were 99 percent sure that they would have known about it if such a specimen had come their way. He went on to speculate, or "gossip," as he said, that the great majority of Hewitt's vast collections had remained behind in Wales, but when the Captain moved to the Bahamas, he took the best and most valuable objects with him. The BTO had been given the former material, and thus never got to see some of Hewitt's best items. There was a suspicion that much of Hewitt's really good stuff in the Bahamas had been disposed of without official record before Spink & Son got to it, if only to avoid taxation, which Hewitt seems to have been particularly keen to avoid. Greenwood also suggested that, whatever the BTO hadn't wanted, had been purchased "by du Pont and is therefore possibly to be found in the Delaware Museum."

Was I any further along in my search? Official sources in the Bahamas were unaware of Hewitt and his duck. It seemed that the Captain had fathered at least one illegitimate child, Major Parry, with his housekeeper. That son, who once lived in Guernsey, is now dead. None of the remaining relatives, including an elderly brother, a son-in-law, and a nephew, were able to tell me where the duck might be. The British Trust for Ornithology got a lot of the Captain's stuffed birds, but didn't get his duck. It may have slipped through the net and found its way to Delaware. If not, it may have been given or sold to someone in the Bahamas with no record. Frustratingly, I was still dealing with rumors; I had no solid evidence that the duck ever really existed.

After a bit of digging, I found that the museum in question is the Delaware Museum of Natural History in Wilmington. Gene Hess of the museum wrote to say that, although they obtained stuffed seabirds and birds-of-paradise from the Hewitt collection, along with a substantial egg collection, the Labrador Duck was not among their holdings. Hess was kind enough to forward my inquiry to Mr. John E. du Pont, founder and former director of the Board of Directors of the Delaware Museum. A response to that inquiry came to me from Mr. du Pont on the beautiful letterhead of Foxcatcher Farms, explaining that the material obtained from the Hewitt collection did not include his Labrador Duck. "In fact, neither I, nor the Delaware Museum of Natural History, has ever owned a Labrador Duck specimen, nor egg, feather, or any part thereof."

I was very impressed that Mr. du Pont, clearly an important man of substantial means, took the time to write to me about something as trivial as a Labrador Duck. A few days later, news crossed my desk to suggest that Mr. du Pont may have plenty of time to answer queries like this. A colleague suggested that I check into John du

Pont using newspaper services available online. Perhaps you already know the story, but it was big news to me. After thirteen days of testimony, and a further week of deliberation, a jury of six men and six women found Mr. du Pont guilty of the murder of Olympic wrestler David Schultz on 26 January 1996. He was sentenced on 13 May 1997 to thirteen to thirty years in prison. The murder of the 1984 Olympic gold medal winner occurred at du Pont's Philadelphia estate and was witnessed by the victim's wife. Psychiatric experts for the defense had argued that du Pont believed that Schultz was a member of an international conspiracy to kill him. John's share of the du Pont empire had an estimated value of $250 million, making him the wealthiest person ever to stand trial for murder in the United States. I suppose his letter concerning the Labrador Duck was written to me from a psychiatric prison. In a curious twist to the story, Mr. du Pont is also famous for owning the rarest and most valuable postage stamp in the world, the Penny Magenta, which he purchased in 1980 for $935,000. That alone might be enough to secure a label of insanity.

IF YOU ARE really sharp, you will have noticed that, to this point, I had no really good evidence that Captain Vivian Hewitt ever actually owned a stuffed Labrador Duck. There is the statement about Hewitt in the 1951 article by Glegg, apparently based on rumor, and an unsubstantiated repetition of those details in Hahn's 1963 summary. It is well within the realm of possibility that I had spent six years chasing an unfounded rumor. At this point, it seemed that different aspects of the Labrador Duck story were beginning to converge.

At the tip of the convergence, I finally made some real progress. When Lisa and I made a whirlwind trip through Scotland, attempting to track down Mr. O'Connell, the owner of a purported Labrador Duck egg, we took the opportunity to drop in on Bob McGowan, senior curator of birds at the National Museum of Scotland in Edinburgh. After showing us some great ornithological treasures, including Great Auk eggs, McGowan continued to show us just how really hospitable Scots are. He took us to the museum's coffee shop, and insisted on treating us. We swapped stories about what brave and honorable men and women ornithologists are, and how wonderful we were to have chosen that path in life. McGowan asked about my work on the songs of birds, and related a story about his wife denying that birds in different regions sang different dialects. I promised to send him some of my publications on exactly that topic. We gradually got around to a discussion of Labrador Ducks.

McGowan said, "You know that one recently came up for sale, do you?" With a wise look on my face, I jumped in to explain that the story about a garage sale duck was just that, a story. No such duck existed, and the story was based on a misremembered conversation. "No," he explained, when I shut up long enough to give him a chance, "I was talking about one that came up for sale in Britain. I think it was owned by a fellow named Hewitt." Asking McGowan for forgiveness while wiping off the decaffeinated coffee that I had spit all over his face, I asked for more details. Fearing a second decaffeinated shower, McGowan explained that whatever further details he had, he didn't think he was at liberty to share. He offered to introduce me to his colleague Andrew Kitchener, principal curator of mammals and birds at the museum, who might be able to tell me a little more.

Lisa and I found Kitchener to be another wonderful and embracing person, but the topic of Hewitt's duck clearly put him a bit on edge. He confirmed that, to the best of his knowledge, it was Hewitt's duck that had recently come up for sale, but he didn't know who the vendor was, and really wasn't at liberty to tell me who had purchased it. He did say he was trying to block the overseas purchaser from taking it out of the country under a British law that decreed that anything of historical or scientific value that had been in the country for more than fifty years required special permit for export. Indeed, if I could help show that the specimen had been in Britain for that period, it would strengthen his case. I promised to send him what I knew of the history of Hewitt's duck, and he promised to be in touch with a friend of a friend who might know more about the purchase of the duck.

True to his word, Kitchener had been in touch with a friend, who had been in touch with a friend. As a result, shortly after arriving back in Canada from Scotland, I received a cryptic email message from a gentleman whom I shall call "Sidney." In it he said: "Ref your request to view the Labrador Duck specimen. The owner is presently out of the country and we are unable to make contact. We will pass on your request when we next see him. Regards 'Sidney.' "

I replied to "Sidney," with gratitude, acknowledging that such matters can sometimes be delicate. I also asked him to pass along to the duck's owner that I am no more than a harmless, if persistent, academic, with no desire to intrude on his privacy. That was something of a lie, of course, because I had every design on the owner's privacy, at least in terms of his newly acquired duck.

I waited for four months for "Sidney" to respond, and then sent him another polite email message, asking for an update. The next day I had a message waiting for me from "Keith," informing me that: " 'Sidney' left our Company in early July this year, and we can find no file on the subject. We act only as forwarding agents to His Excellency. I feel it is best that you make your approach direct to the Embassy of Qatar, as we have no interest or input in this matter. Regards 'Keith.' "

"His Excellency"? "Qatar"? What in hell was all of this about? "Keith" clearly wasn't kidding when he said that they had no further interest in the matter. I sent additional messages begging for more information, but they all went unanswered; I couldn't get a peep out of them. It was time to resort to that least reliable of all sources of information, the Internet. The company that employed "Sidney" and "Keith" was nowhere to be found on the Internet. Frankly, anything that manages to keep itself hidden from the invasive tendrils of the Internet frightens me, and so I'll leave the company unnamed, and will never mention them ever again in the hope that they won't hunt me down and kill me.

I was, however, able to find that Qatar, an Islamic country with only a few hundred thousand citizens, is now one of the richest nations in the world, largely on the basis of its vast oil and gas reserves. The Al Thani family has ruled Qatar for about 1,500 years. Since 1995, the Emir of Qatar has been Sheikh Hamad Bin-Khalifah Al Thani, who took control of the country from his father, Sheikh Khalifah Bin-Hamad Al Thani, who was out of town at the time. I suppose that shows what can happen when the children start saying: "Gee, Dad, we don't need a babysitter!" Could it be that the Emir had some odd interest in Labrador Ducks and enough pocket change to get one?

I wrote again to Kitchener in Edinburgh, suggesting that the cat was now almost completely out of the bag, and asked him to open the bag just a little more to allow the cat to escape so that it might resume ripping up the living room drapes. Was Sheikh Hamad Bin-Khalifah Al Thani the man with the duck? He responded by saying that, to the best of his knowledge, the duck had been purchased by Sheikh Saud bin Mohd. bin Ali Al-Thani, but that he did not know the bird's current location.

The storyline was getting too complex for a Harlequin Romance, but at least I had evidence that the specimen really existed. According to du Pont, the Hewitt duck never got as far as Delaware. Rumor suggested that the duck may have been purchased by a Sheikh in Qatar, who was trying to get it out of Great Britain despite protestations; the whole affair was all very hush-hush, making my investigation all the more difficult.

But for once, I knew a little more about this particular Labrador Duck than someone else. Despite Andrew Kitchener's protestations about the need to keep the duck in Britain, I had been tipped off that it had already been taken to Qatar in the only place that a customs agent couldn't look—a diplomatic bag. I was now in the deep end of the pool and I couldn't swim. I wasn't even wearing water-wings. Or a swimsuit. Knowing who owns the duck was no good when I didn't know how to contact him, or even if it was possible to contact him. I am a lowly biology professor. I do not run in the sort of circle that allows me to call up a chum and say, "Listen, Reginald old buddy—you play cricket with Sheikh Saud of Qatar, don't you? Do you think you could ring His Excellency up and ask him round for gin-and-tonics so that I could have a little chinwag with him? It's just that he seems to have stumbled across a little duck thingy that I am interested in . . ."

If you are curious, and if you never throw away your old copies of *National Geographic* magazine, you can see a picture of Sheikh Saud.

In an article about Qatar's move to modernize itself, a two-page photograph of Sheikh Saud inspecting his oilfields from the comfort of his silver convertible BMW Z8 appears on pages 84 and 85 of the March 2003 issue. With both hands on the wheel, but with his left arm resting in a jaunty fashion on the door frame, sporting ultra-cool sunglasses, with a cheap gift-bag at his right shoulder, he is the model of a man in control of his destiny.

I couldn't immediately think of any way to make forward progress in finding Hewitt's duck, so I decided to try approaching the problem from the other end by going after Mr. R.L.E. Ford, who had discovered the duck in Kent somewhere around 1947. Fortunately for me, the company of Watkins and Doncaster, in operation since 1874, still exists, and is operated by Mr. Ford's son, Robin. He was able to tell me that the senior Mr. Ford had passed away four years earlier, and that company records from the Labrador Duck era had been destroyed when the company moved locations in either 1956 or 1973. This didn't leave much for me to follow up on.

Surely there had to be a way to find the home in Kent where the Labrador Duck had been found. I contacted Kenny Everett, an English gentleman with a fondness for historical taxidermy, who told me that many of the great natural history collections in Kent had been destroyed in World War II, but suggested that I try to contact Errol Fuller, a noted British author, and a serious collector of natural history artifacts with a "tremendous collection."

I knew about Errol from his wonderful books on Great Auks and other extinct birds, and was destined to become good chums and would later go on Labrador Duck adventures with him, but to this point I had no reason to contact him. In doing the research for his books, Errol had corresponded with folks in many of the same museums as I, which, I felt, might give us some common ground. So, briefly describing my search for Hewitt's Labrador Duck, I told him that, while thinking of a way to find out about the Sheikh, I was keen to find the house in Kent where Ford had first discovered it. His response could not have been more welcome.

Errol did not know in which house in Kent Ford had made his fantastic discovery, but he knew the current location of Hewitt's Labrador Duck, and knew the owner personally. Imagine me making choking noises while swallowing my tongue. At that time, however, Errol wasn't able to give me all of the details, having promised discretion and secrecy. As a writer, he fully appreciated the frustration of being told "I know what you want to know, but I can't tell you," and he promised to tell me all, eventually. I wrote back to say that Lisa and I were going to be in England later that year, and offered to buy him a bathtubful of beer at his local pub in the hopes of loosening his lips. He agreed.

A few months later, Errol met Lisa and me at a train station near his home, and we became immediate friends. He showed us around the collection of natural history artifacts that occupy his home. We talked about England's exit from the World Cup soccer tournament in the second round, and about the £10 wager he had made on Senegal to win that day's game at 16 to 1 odds. (Senegal lost.) Over a curry at a restaurant close to his home, Errol spilled the beans about Hewitt's duck.

Captain Vivian Hewitt had not taken all of his collections to the Bahamas. The sheer bulk of the material he had collected over a lifetime would have made this impossible. Even some of his most valued possessions had remained behind in Wales. Hewitt's "housekeeper," Mrs. E. M. Parry, was a lot more to the Captain than that, and bore him three children, who referred to him as "Dad" around the house and "Captain" in the street. Of those children, only Jack Parry was still alive, but very elderly, and not necessarily watching television on the same channel as the rest of us.

On Hewitt's death, his children divided up the Captain's treasures. Spink & Son sold the items with obvious value, such as the Captain's enormous coin collection. There were something like 100,000 bird skins and 500,000 eggs in the collection at the time. Some of these, like the Great Auks, were of obvious value, and Spink & Son also dealt with those. Much of the remainder of the bird collection remained at the family house in Anglesey, Wales, and son Jack got that part of the Captain's collection. Somewhere around 1968, not so many years after Captain Hewitt's death, someone at the Royal Society for the Protection of Birds told Jack that it was illegal to sell the specimens. This was entirely untrue, but it so upset Jack that he set his mind to tip the whole collection over a cliff into the ocean, and so end his (and everyone else's) association with the specimens.

Luckily, a seller of antique bird books, David Wilson, heard about Jack's intentions, and drove from his home near Tring to Anglesey to try to change Jack's mind. Jack wanted no further responsibility for the birds, and offered to give Wilson the whole collection on the spot. It was agreed that Wilson would return as quickly as possible with a small fleet of lorries to take the collection away, with the intention of turning it over to the British Trust for Ornithology. Parry and Wilson shook hands on the porch, and the deal was done. After some dickering, the collection was placed in the basement of the Walter Rothschild Museum in Tring, where the British Trust for Ornithology gradually sold off most of the items of value. A number of museums helped themselves to other bits and pieces, and finally John E. du Pont bought everything else, and took it to the museum in Delaware.

Wilson had kept a very small number of items from Hewitt's collection, including Hewitt's personal papers, a stuffed Snowy Owl, a model of a Great Auk, and four or five other birds. It was a pretty modest haul, considering that Wilson technically owned the whole lot. Errol had met Wilson while working on his Great Auk book, and had always found him to be helpful and kind. Wilson knew that Errol had a great interest in natural history artifacts, so when he decided to sell what remained of the Hewitt material that had been in his possession for nearly a third of a century, he asked for Errol's help in establishing prices. Wilson knew that there are a lot of unscrupulous dealers who would have been very happy to pay him a pittance to part with the items. Errol agreed to help out; having seen most of the items in Wilson's home before, he knew it wouldn't be too onerous a task.

But then Wilson took Errol up to his attic, and showed him my holy grail. It was Hewitt's Labrador Duck. Errol, always keen to get his hands on precious artifacts, bought the duck, and took it home. Errol kept it for two months before he found himself short of cash, at which point he put it up for sale with a London art dealer, who sold the duck to Sheikh Saud of Qatar, a man with a thirst for acquiring beautiful and rare objects, for whom money is not the greatest limitation in life.

It seemed that there was no way for me to find out which country home in Kent had housed the duck when Ford found it. However, at this point I knew quite a lot about the comings and goings of the specimen. For instance, I knew that the duck had not followed Captain Hewitt to the Bahamas, but had remained behind in Wales with one of the Captain's sons, Jack Parry. When the Captain died, someone had convinced Parry that having a collection of natural history artifacts was illegal. In the nick of time, David Wilson had stepped in, convincing Parry not to destroy the collection, but to pass it along. Wilson saw that suitable recipients got the collection, but retained the Labrador Duck for himself. This he sold many years later to Errol Fuller. Shortly after, it was purchased by Sheikh Saud.

Now, of course, it was a matter of finding a way to see this, my fifty-fifth and final duck. This was going to be tricky. There was talk of the Sheikh, head of Qatar's National Council for Arts, Heritage and Culture, establishing a national museum which would house an astonishing collection of natural history rarities. To this point, the museum was more of an idea than an institution, and it was unlikely to be open to the sort of request I had made of all the other museums with Labrador Ducks. The Sheikh was one of the richest men in one of the world's richest countries. What could I offer him in order to be given the opportunity to see his duck? It really seemed as though I would pull up one bird shy of my goal to see all of the world's Labrador Ducks.

And then, on a Monday afternoon in August, fully nine years, one month, and six days after my search began, "the call" came through. It was from Errol Fuller. He wanted to know if I could fly to London immediately. Captain Hewitt's duck was back in Great Britain, but the window of opportunity to see it was extremely narrow. Lisa leapt into action, and secured me a seat on a flight arriving at Gatwick at 10:50 the next morning.

Errol and his son, Frankie, were waiting for me at Gatwick. As we drove off, Errol brought me up to date. The duck had been taken from England to Qatar in a fashion that, if not strictly illegal, was certainly not done by the books. The Sheikh, in an apparently random whim, had decided that his Labrador Duck was to be shipped back to England so that the proper paperwork could be completed for its removal to Qatar. Since the duck had been residing in Qatar for two years, and no one had any way of forcing its extradition to England, this seemed a rather odd move. Furthermore, there was a degree of urgency to complete the paperwork, as the duck was going to be one of the star attractions in an exhibition in Qatar ten weeks hence. Indeed, the paperwork might come in the form of a temporary export license to facilitate the duck's appearance in the exhibition, although the chances of the duck's returning to England afterwards seemed extremely remote. Regardless, the duck was residing at an art gallery near Piccadilly and had to be taken to a firm of international art shippers in Vauxhall, where it would be held until all of the necessary paperwork would be completed, at which time it would be shipped back to Qatar. Errol was well known to the gallery owners who had taken him up on his offer to transport the duck to the shippers. The interval between the art gallery and the shipper was to be my opportunity to examine the duck. The plan was to take the duck to Errol's home so that I could examine it at leisure. We would give it to the shippers when Errol took me back to the airport.

At the art gallery, on a street lined by an assortment of posh art galleries, we were buzzed in. Errol thought it prudent not to mention my specific interest in the duck, and so I was introduced as his colleague. I was expecting the duck to be brought to us in a stout and professionally constructed shipping container, with lots of wood and big brass screws. I was surprised when we were brought a reused cardboard box sealed with masking tape and secured with a bit of string. It sported two small red "fragile" stickers, and the words LABRADOR DUCK were scribbled on the side in felt pen. Frankly, my Christmas ornaments reside in a more secure box.

Labrador Duck 55

A "cursed" duck, and the final specimen in my quest to see every Labrador Duck in the world. Having changed hands many times, it is now in the collection of Sheikh Saud of Qatar.

When we got to Errol's home, he cleared a bit of space at his kitchen table, and I set the box down. For some reason, I wasn't in a terrible hurry to get on with my examination. Perhaps it was because we were suffering through a very hot and humid day, and I needed a breather. More likely it was because the box contained an object that I had been searching for for almost a decade, and it was to be the last new Labrador Duck I would ever see. This was it. Once I opened the box and poked at it a bit, the adventure would be over.

Frankie watched me as I cut the string, pulled off the masking tape, and reflected the box's flaps. At first all I could see was crumpled tissue paper, but as that came away, the duck revealed itself, and my breath rushed into my chest with a whoosh.

I had seen a scanned image of a photocopy of a fifty-year-old, black-and-white photograph of the duck being examined by Richard Ford before selling it to Captain Hewitt, and from what detail remained after so much time and copying, it seemed a truly lovely specimen. Errol had warned me that the duck had been restuffed about 18 months earlier. This had been a really dangerous move; in taking apart a specimen this old, even the most skilled taxidermist might have been left with nothing more than a pile of feathers. However, Errol had assured me that a good job had been done. My first impression was of a very artistic interpretation. When it was first found, the duck had seemed a little too alert, but the new taxidermist had given it a more relaxed look. The neck was more bent, bringing the head closer to the shoulders, and the eyes were ever so slightly closed, as though the drake was getting ready to nod off. Originally, the duck had been mounted on a base covered in pebbles and seaweed, suggesting, I suppose, the seaside where the duck had lived. This had been replaced by a simple wooden slat, 18 by 19 by 2 cm, painted black, which allowed viewers their own interpretation of the duck's circumstances.

So far so good. Regrettably, the fellow who restuffed the duck had taken it upon himself to paint the legs, toes, and webs a uniform gray-blue. Even worse, he had painted the bill. The distal half was black, and the proximal half was mustardy yellow, overlain along the midline by a broad triangle of gray-blue. To give the taxidermist credit, he was thorough, having painted even the underside of the bill in the same mustard-yellow and battleship gray.

So I made my measurements, and drew little pictures in my notebook, and used a magnifying lens to examine wear to the tail and wing feathers. And as I poked and peered, the sun broke through the persistent British clouds, and a shaft muscled its way through Errol's kitchen window, and fell on the duck. Something strange happened. In the afternoon sun, a few patches of particularly black feathers glowed an iridescent green. I was ready to blame my dirty contact lenses, but when I called Errol over, he said that he could see the green sheen too. Some birds have patches of iridescent green feathers, including some ducks. Was this a feature of Labrador Ducks that I had somehow missed? If the sun hadn't come out, I wouldn't have spotted it this time. Could all of the authors who described Labrador Ducks in life so long ago have missed this iridescence? There was certainly no green hue in Audubon's painting. I looked, and turned my head, and blinked, and looked again, and convinced myself that the green iridescence was the result of a little subterfuge. I believe when the taxidermist was restuffing the duck, he found himself short of feathered skin in spots, and used bits from another duck, possibly a Mallard, to fill in the gaps.

And there we have it. Some sixty years ago, this Labrador Duck was owned by someone in a country house in Kent, then by naturalist Richard Ford, by Captain Vivian Hewitt, by Hewitt's son, Jack Parry, by bookseller David Wilson, and by my chum Errol Fuller, after which it came into the possession of an art dealer, who sold it to Sheikh Saud of Qatar. And I suppose that is the end of my story. Barring some unforeseen discovery, I can claim to have seen absolutely every stuffed Labrador Duck in the world, and no one else could make that claim, and I am sure that no one ever will. This left only a quick trip to the shippers to exchange the duck for a receipt numbered 4513/04.

Of the owners of this duck, a pretty good chunk are known to be or suspected of being dead, or well on their way. Many other protagonists in the story are dead. To be fair, an historian friend of mine pointed out that the same can be said of any very old artifact, but part of me thinks that this particular duck is not particularly charmed. Could it be cursed? Let's wish Sheikh Saud a very long and happy life.

I TOLD YOU earlier that I wouldn't reveal the end of the story of *The Maltese Falcon*. Upon further reflection, this seems to be a tease. After all, your local video store is bound to have hundreds of copies of *Spider-Man* and *Harry Potter,* but the odds are about three to one that the clerk has never heard of *The Maltese Falcon*, and would probably confuse Humphrey Bogart with Harrison Ford.

At the end of *The Maltese Falcon*, the statuette that the crew have been following turns out to be a fake, and Caspar Gutman leaves to continue his search for the real one. In the movie, Sam Spade calls the police to put them on to Gutman as a murderer, but Gutman escapes police custody, and, presumably, continues his quest. They do capture the young lady and charge her with murder. The end of the book is a little more gruesome than the movie. Gutman dies in a shoot-out with police, and so never gets the falcon statuette that he has been following for seventeen years. I searched for this single specimen of the Labrador Duck for more than nine years. Unlike Caspar Gutman, I eventually found it, and didn't die in the chase.

Epilogue

And that was that. I had been around and around the world with strange and wonderful traveling companions. I had examined fifty-five stuffed Labrador Ducks in thirty cities, and sampled nine eggs that later proved to have nothing to do with the birds I was after. From standing on Audubon's hillside in Labrador to measuring my last duck, my adventures had taken me four years, nine months, and eighteen days. I had, as far as I could tell, seen every surviving stuffed Labrador Duck in the world. The end of the quest left me with a sense of elation at having completed a task that no one had ever attempted before, and that no one would ever bother to repeat. After all, twelve people have stood on the Moon, but I was the only person to have seen every Labrador Duck. But I also felt a sense of deflation that the journey was over.

To be completely forthcoming, there were several Labrador Duck specimens that I couldn't account for, and, as friends have pointed out again and again, it is possible that new specimens remain to be discovered. How confident am I that I found them all? So confident that I will pay a reward of $10,000 to the first person who can direct me to a genuine stuffed Labrador Duck that I have not seen and described in this book.* I don't want to buy the duck; I just want to ex-

* I should make clear that this offer comes from me alone and not from my publishers or anyone else involved in bringing you this book. For more information, and for the full terms and conditions of the reward, visit my website, www.glenchilton.com.

amine it. After I have verified its legitimacy, you get the money. There is no point in trying to fake me out with a duck re-created from bits of other birds; I've seen it all before.

If you choose to embark on your own Labrador Duck quest, you might want to start with the specimens that may have been destroyed by wartime bombing in Liverpool, Amiens, and Mainz. Perhaps you know the reprobate that stole the duck from the American Museum of Natural History in New York; perhaps you *are* the reprobate. Altenburg probably never had a Labrador Duck, but it could be worth a peek if you are in the area.

These are all long shots, but there are other possibilities. In April of 1962, Paul Hahn of the Royal Ontario Museum received a response to his questionnaire about stuffed extinct birds from Everett F. Greaton, a consultant for the Recreation Department of the Department of Economic Development at the State House in Augusta, Maine. Greaton indicated that the bird section of the museum at the State House had a stuffed Labrador Duck. Hahn didn't include that specimen in his catalogue, and forty years later, Kris Weeks Oliveri, Volunteer Coordinator at the Maine State House Museum, assured me that there is no sign of a Labrador Duck in their collection. Maybe it is sitting on a shelf in the Governor's office.

I'll give you one more lead, and then you are on your own. In November of 1844, Colonel Nicolas Pike shot a drake Labrador Duck at the mouth of the Ipswich River at the south end of Plum Island, New York. History doesn't say whether or not the duck was doing anything to provoke the colonel. Perhaps Pike just really, really hated ducks. The bird was stuffed by John Akhurst, given to the Long Island Historical Society (now the Brooklyn Historical Society), and eventually deposited in the Brooklyn Museum. The Brooklyn Museum narrowed its focus in the 1930s and 1940s, and dispensed with its natural history collection.

At my request, Deborah Wythe of the Brooklyn Museum did some digging. Between them, Charles Schroth and Charles O'Brien at the American Museum of Natural History received from the Brooklyn Museum fifty-eight cartons of bird study skins in August, 1935. If birds are packaged like cigarettes, twenty-five to a pack and eight packs to a carton, we are talking about roughly 1,600 specimens.

No mention was made of mounted specimens in general, nor about the Brooklyn Labrador Duck in particular. I have already examined every Labrador Duck specimen at the AMNH and it isn't there. At Wythe's suggestion, I tried the Brooklyn Children's Museum, which apparently got many natural history specimens without much ceremony or documentation. Nancy Paine, who has the delightful job of Chief Curator at the Children's Museum, informed me that I had hit another dead end. John Hubbard told me that the Bailey-Laidlaw Collection at Virginia Tech in Blacksburgh had received a chunk of the Brooklyn Museum's collection, but Curt Adkisson, current curator, was able to tell me they do not have, and never did have, a Labrador Duck. It must be out there somewhere.

Good luck in your quest.